Chemistry Lab Investigations

For AP & IB Students

The book provides coverage of the essential lab topics of the AP and IB Chemistry courses. Each lab investigation is well-structured with an introduction, lab concepts, procedure, execution, results, analysis, and conclusion.

- **Saurya Singh**

Kalisey Academy Publication

Kalisey Series

<u>Delivering the highest quality academic materials</u>

About the Author

Saurya Singh was born on 17.Nov.2004 in Frankfurt am Main. Right from his childhood, he enjoyed being involved in math-puzzles, Mathletics, Abacus, and science lab investigation. He had great passion for mathematics and science performing well in advanced level at a quite early age.

Saurya also conducts workshops of math and science and enjoys being involved in blogs. In his books, blogs and workshop, he encourages the enquiry based learning and promotes the transdisciplinary approach to understanding. His approach stimulates the interest and nurtures the skills in math and science.

Among extra-curricular activities, Saurya is very much active in sports and music. He is a fantastic badminton player. His other activities include swimming, martial-arts and biking. Saurya loves playing violin.

Colophon

Title: Chemistry Lab Investigations – For AP & IB Students

Description: The book presents essential chemistry lab investigations for AP & IB students

Author: Saurya Singh

Series: Kalisey Series

Publisher: Kalisey Academy Publication

Book Presenter: Author & publisher

Copyright: Author & Publisher

Website: www.kalisey-softek.com/chemlab.html

Contact: info@kalisey-softek.com

17.Jan.2021 - First release;

Wherever the concept and information has been used from other resources like standard rules/frameworks/methods etc. and the author was aware of reference, the reference of original source has been mentioned in the book.

The author, host, publishers or any other entities involved in preparing and publishing this book make no expressed or implied warranty of any kind and assume no responsibility for errors or omission or incompleteness in the book. No liability is assumed for incidental or consequential damages arising out of use of the concepts or information contained herein.

> **Dedicated to my parents, teachers & my school – Frankfurt International School**

Copyright © 2021 Author & Publisher

All rights reserved.

Copyright © 2021, Author & Publisher. All rights reserved. No part of this publication may be reproduced in any form by print, photo print, microfilm or any other means without written permission.

For any further enquiries about the book and content, contact: info@kalisey-softek.com

Contents

1. LAB 1: METALLIC, IONIC, COVALENT, AND MOLECULAR SOLIDS 9
2. LAB 2: MOLE RATIO 21
3. LAB 3: PAPER CHROMATOGRAPHY 35
4. LAB 4: TYPES OF CHEMICAL REACTIONS 51
5. LAB 5: COLORIMETRY 73
6. LAB 6: GRAVIMETRIC ANALYSIS 97
7. LAB 7: BOND AND MOLECULAR POLARITY 115
8. LAB 8: INTERMOLECULAR FORCES 141
9. LAB 9: REVERSIBLE REACTIONS 155
10. LAB 10: ACID-BASE TITRATION, BUFFER SOLUTION 171
11. LAB 11: REDOX TITRATION 199
12. LAB 12: ELECTROCHEMISTRY – GALVANIC CELLS 217
13. INDEX 231

Preface

Right after completing my grade 9 in the year 2020, I started my CTY AP Intensive Chemistry course under Dr. Lev Ryzhkov, who inspired and helped me reach new heights in chemistry. I truly enjoyed the course and especially the lab investigations required for the AP Chemistry course.

I was very much interested to perform the experiments at home with my Chemistry Lab-Kit and had a lot of fun investigating the principles of the science behind them. Some of my key lab investigations, which are immensely fascinating include:

- Identifying the types of solids and the forces in action by physical properties.
- Investigating the mole ratio in a chemical reaction.
- Separating the solutes from a mixture using chromatography.
- Finding out the amount of phosphate in plant food.
- Simulating and analyzing the bond polarity, partial charges, and electrostatic forces using electronegativity.
- Investigating the reversible reaction and applied Le Chatelier's principle.
- Performing acid-base titration and investigating the properties of the buffer solution.
- Finding oxidation states using redox titration.
- Constructing a galvanic cell and determining the cell voltage.

One of the core intellectual aspects of my lab investigations was to connect the principles of science in the context of real-world problems involving a transdisciplinary approach.

Apart from learning from my lab investigations, completing the labs for the AP course was quite involving. Having invested so much effort, I wished to share my work with others. I altered, modified, and consolidated my work of research to publish this book.

I convey my special thanks to my school teachers Ms. Zina Sawabini and Ms. Jessica Angelidis, Counselors & University Advisors in FIS, for registering, proctoring, and supporting me to complete my CTY courses and AP exams.

The book provides coverage of the essential lab topics of the AP and IB Chemistry courses. Each lab investigation is well-structured with an introduction, lab concepts, procedure, execution, results, analysis, and conclusion.

Have fun performing chemistry experiments!

- ***Saurya Singh*** (Frankfurt am Main, *14.Jan.2021*)

1 Lab 1: Metallic, Ionic, Covalent, and Molecular Solids

Goals
To study the differences between metallic, ionic, covalent network, and molecular solids.
To identify the types of solids from physical characteristics.
To investigate the conductivity of metallic, ionic, covalent network, and molecular bonds

Introduction
A solid can be composed of atoms, ions, or molecules. Physical and chemical properties can be differentiated by the structure and the arrangements of atoms, ions, and molecules and the forces between them.

There are four types of crystalline solids:

1-Metallic solids: Metallic solids are made up of metal atoms that are held together by metallic bonds. These solids are characterized by high melting points, can range from soft and malleable to very hard, and are good conductors of electricity.

2-Ionic solids: Ionic solids are made up of positive and negative ions and held together by electrostatic attractions. These solids are characterized by very high melting points and brittleness and are poor conductors in the solid-state. An example of an ionic solid is table salt, NaCl.

3-Covalent-network (also called atomic) solids: Covalent network solids are made up of atoms connected by covalent bonds; the intermolecular forces are covalent bonds as well. These solids are characterized as being very hard with very high melting points and being poor conductors. Examples of this type of solid are allotropes of pure carbon; diamond, graphite (good conductor), and fullerenes.

4-Molecular solids: Molecular solids are made up of molecules held together by intermolecular forces like London dispersion forces, dipole-dipole forces, or hydrogen bonds. These solids are characterized by low melting points and flexibility and are poor conductors. An example of a molecular solid is sucrose. The atoms within the molecule are linked together by sharing of electrons through covalent bonds, which implies that the forces acting between atoms within the molecule (intramolecular) are much stronger than those acting between molecules (intermolecular).

Metallic substances are characterized by the delocalization of electrons over the entire solid, which makes them a good conductor besides bonding the metal atoms strong enough to be generally in the state of solid at room temperature.

Chemical compounds are combinations of atoms held together by chemical bonds, which are of two basic types – ionic and covalent.

Ionic bonds result when one or more electrons from one atom are transferred to another atom. The atoms form positive and negative ions which attract to each other. Ionic compounds are composed of large numbers of positive ions (cations) and negative ions (anions) so that the amount of positive charge equals the negative charge. The ions locked in the rigid lattice do not allow free movement of electrons, so solid ionic substances do not conduct electricity, but they do conduct when the ionic substances are dissolved in solution and the ions are mobile.

In covalent bonds, the electrons are shared between two bonded atoms. A molecule is a group of atoms that are bonded together by covalent bonds. Molecular compounds are composed of these molecules. However, the intermolecular forces that condense from a gas to a liquid or solid are much weaker than covalent bonds. Since there are no free electrons, molecular substances do not conduct electricity in solids. Atoms that form extensive covalent bonds are in the category of covalent network substances. These substances have a very high melting point, and are also characterized by physical characteristics of being malleable and ductile. Covalent network solids may be good electrical conductors or insulators.

Molecular solids are solids that are essentially collections of molecules held together by intermolecular forces, but not by strong bonds (metallic, covalent, or ionic). The forces holding the solids together are much weaker than for other types of solids.

The physical properties of a substance such as melting point, solubility in water, and conductivity of an aqueous solution express a lot about the type of particles in a compound.

Hypothesis and variables

Hypothesis: Molecular compounds are bad conductors, whereas covalent compounds are good conductors of electricity. The ionic compounds are bad conductors in solid-state, but become good conductors in solution due to the availability of free ions.

Independent Variable: Substances to be tested for conductivity.

Dependent Variable: The strength of brightness of LED (bright red light, dim red light, no light).

Controlled Variable: Room temperature.

Materials
- Ascorbic acid
- Palmitic acid
- Paraffin
- Aluminum foil
- Distilled water
- Pencil lead
- Salt (NaCl)
- Sugar (Sucrose)
- Wax paper
- Spatula

- Test tube, 12 x 75 mm
- Conductivity tester with a 9-volt battery
- Digital balance
- Graduated pipette
- Wire gauze
- Heat source
- Paper towel

Procedure

Part-1: Testing the conductivity of substances at room temperature
1. Six substances were tested for conductivity at room temperature
2. The substances were aluminum foil, water, pencil lead (graphite), sucrose, ascorbic acid, and sodium chloride.
3. The 9-volt battery was attached to the conductivity tester, and both the electrodes were touched to different substances for testing conductivity.
4. The strength of brightness of LED was marked with bright red light with good, for dim red light with poor, for no light as non-conducting.
5. The data were recorded in Table-1-1

Part-2: Testing conductivity of solutions
1. Three solutions of substances were tested for conductivity
2. The substances were sucrose solution, ascorbic acid solution, and sodium chloride solution
3. The three different solutions were prepared by adding a pinch of each substance into different water beads.
4. The aqueous solutions were tested for conductivity and data were recorded in Table-1-2.

Part-3: Testing conductivity of pure liquids
1. Four substances were tested for conductivity in the state of liquid
2. The substances were salt, palmitic acid, paraffin, and sucrose.

3. These substances were kept in different quadrants of a circular aluminum foil and heated on a stovetop till the substances melted completely.
4. It was found that salt could not be melted due to its higher melting point than aluminum, so it could not be tested.
5. The melted substances were tested for conductivity and data were recorded in Table-1-3.

Results

Data recorded in Table-1-1 for the test of conductivity of different substances at room temperature.

Substances	Conductivity
Aluminum	Good
Water	Bad
Graphite	Good
Sucrose	Bad
Ascorbic acid	Bad
Sodium Chloride	Bad

Table-1-1: Conductivity of substance at room temperature

Data recorded in Table-1-2 for the test of conductivity of different solutions.

Solutions	Conductivity
Sucrose solution	Bad
Ascorbic acid solution	Bad
Sodium Chloride solution	Good

Table-1-2: Conductivity of different solutions

Data recorded in Table-1-3 for the test of conductivity of different substances in a melted state.

Solutions	Conductivity
Sodium Chloride	Could not be melted, conductivity test not done
Melted Palmitic acid	Poor
Melted Paraffin	Bad
Melted Sucrose	Bad

Table-1-3: Conductivity of melted substances

Analysis

In the investigation of part-1, the data is recorded in Table-1-1. In a metal like Aluminum, the valence electrons are loosely held. They leave their own metal atoms, forming a sea of electrons surrounding the metal cations in the solid. The electrons are free to move throughout this electron sea. Thus, metals are good conductors of electricity.

There are three allotropes of pure carbon: graphite, diamond, and fullerenes.

Graphite, although also only made up of carbon atoms, is the only non-metal that can conduct electricity. This is because each carbon is trigonal planar and only 3 of the available valence electrons form covalent bonds (σ bonds) leaving 1 spare electron, which then becomes a delocalized sea of electrons loosely bonding (π bond) the layers together. This delocalized electron is no longer associated with one particular carbon atom and it is able to move freely between the carbon layers of graphite. These free electrons make graphite a good conductor of electricity.

In the other allotropes of pure carbon diamond and fullerenes, the carbon atoms are joined by strong covalent bonds. Every atom in a diamond or fullerenes is bonded to its neighbors by four strong covalent bonds, leaving no free electrons and no ions. Hence, diamond or fullerenes does not conduct electricity.

Sucrose is a molecular substance that contains molecules held in place with the London Dispersion Force, a type of intermolecular interaction. It does not contain free electrons, thus do not conduct electricity.

Ascorbic acid is a covalently bonded substance that has hydrogen ions, which act as charge carrier, hence conduct electricity.

Salt is an ionic compound. The ions in the crystal cannot move, so NaCl in a solid-state does not conduct electricity.

In the investigation of part-2, data was recorded in Table-1-2.
Dissolved sucrose crystals overcome the intermolecular forces. However, it does not break the covalent bonds within individual

molecules. As a result, sucrose won't conduct electricity even if dissolved in the solution.

Ascorbic acid is a covalently bonded substance that has hydrogen ions when dissolved in water, which act as charge carriers, hence conduct electricity.

In the salt solution, the sodium and chlorine ions are pulled apart, so they can floating freely. These ions are what carry electricity through the water with an electric current.

In the investigation of part-3, data was recorded in Table-1-3. Melted paraffin, and melted sucrose are not good conductors of electricity as all of these belong to the category of weak acids. Melted palmitic acid is slightly acidic with high pKa like weak acids, so it is a poor conductor.

Substances	Classification	Justification 1	Justification 2
Aluminum	Metallic	Good conductor at room temperature	Solid, ductile, malleable
Ascorbic acid	Molecular	Bad conductor at room temperature	Low melting point; Crystal form; Not hard like ionic or metallic substances;

Graphite	Covalent network	Good conductor	Solid-state at room temperature; Hard and brittle;
Paraffin	Molecular	Bad conductor in melted state	Low melting point
Palmitic acid	Molecular	Bad conductor at room temperature	Low melting point; Not strong like ionic or metallic substances;
Sodium chloride	Ionic	Bad conductor in solid-state, but good conductor in solution	High melting point; Crystal form at room temperature;
Sucrose	Molecular	Bad conductor of electricity	Low melting point
Water	Molecular	Bad conductor of electricity	Liquid state at room temperature

Table-1-4: Classification of substances with justification

The atoms of HCl, i.e. H and Cl atoms, have a covalent bond. So, pure HCl is a molecular substance. Pure HCl therefore does not conduct electricity. When HCl is dissolved in water forming hydrochloric acid, it readily dissociates into H+ and Cl- ions. That is why hydrochloric acid is a good conductor of electricity. So HCl solution (hydrochloric acid) is an ionic substance. Hence, HCl can be classified as either molecular (polar covalent) or ionic depending on the concentration of the molecule.

Error-Mitigation

a- The conductivity test should be carried out at a normal room temperature, because temperature affects the conductivity.

b- The 9-volt battery attached to the conductivity tester should be new or assured to be in working condition.

c- The electrodes of the conductivity tester should be clean.

d- The aluminum foil being heated on a stovetop should get uniform and consistent heat all around to melt all the substances kept on it uniformly.

e- The strength of the brightness of LED should be observed in a relatively dark place. Too much bright place might cause erroneous data for the strength of the brightness of the LED.

Conclusion

In a metal like Aluminum, the valence electrons are loosely held, so they are a good conductor of electricity. Metals are generally very hard, ductile, malleable, and shiny.

The molecular substance does not have free-flowing electrons, as the bonds are formed due to the sharing of electrons, so they do not conduct electricity. Besides that their melting point is significantly less than ionic or metallic substances. These substances are generally bad conductors of electricity in all of the forms, solid form, melted form, or in solution form.

In Ionic substance, the type of bond occurs in salts having one metal and one non-metal element. They are soluble and dissociative into their ions, displaying strong conductivity in solutions. The salt is a bad conductor in solid form, but is a good conductor when dissolved in water.

The covalent network substance graphite is observed as a good conductor of electricity. Graphite, an allotrope of carbon, is a very good conductor of electricity due to the mobility of the electrons in its outer valence shells.

It was hypothesized at the beginning of this investigation that molecular compounds are bad conductor of electricity, covalent compounds are good conductor of electricity, and the ionic compounds are bad conductor in solid-state but become a good conductor in solution due to the availability of free ions, which turns out to be true as observed in Table-1-1, Table-1-2, and Table-1-3.

The real-world connection of this investigation is that the behavior of substances is exhibited by the chemical constitution and atomic arrangement of substances.

Lab 2: Mole Ratio

Goals
To study the mole ratio in a compound, and mole ratio relationship in chemical reactions
To connect the number of particles, moles, mass, and volume of reactants to one another in a chemical reaction

Introduction
The atoms are chemical elements, which are fundamental building materials of matter. These atoms retain their identity in chemical reactions. Changes in matter involve the rearrangement of atoms and the transfer of electrons.

Generally, for a chemical reaction, the mole ratio of reactants, and the mole ratio of each reactant and product are fixed and well defined. Chemical reactions follow the law of conservation of mass, law of constant proportions, and law of multiple proportions, thus, the reactants and product maintain a mole ratio relationship. But there are many chemical reactions occurring in certain conditions, where different mole ratios are encountered even when the reactants are the same. With a varying mole ratio of reactants, different precipitates are formed. The mole ratio may be determined by examining the coefficients in front of formulas in a balanced chemical equation.

Precipitation reactions are a type of double replacement reaction in which the pattern of reaction is $AX + BY \rightarrow AY + BX$. Here A and B are cations, and X and Y are anions. The cations and anions switch partners. These reactions have a product that is a precipitate.

Stoichiometry is used to balance equations and to specify the mole ratio (or molar proportions) of elements in a compound. Using stoichiometric calculations, the amount of one or more reactants required to produce a particular amount of product can be determined and vice-versa.

In this lab, the amount of precipitate is observed and the concepts of mole ratio are verified.

Hypothesis and variables

Hypothesis: Maximum precipitation in a well shows the actual mole ratio of reactants in a chemical reaction.

Independent Variable: Mixing different volumes of reactants.

Dependent Variable: Amount of precipitation in reacting wells.

Controlled Variable: Concentration of reactants, temperature.

Materials

- 0.1 M Potassium iodide KI
- 0.1 M Sodium oxalate $Na_2C_2O_4$
- 0.1 M Calcium nitrate $Ca(NO_3)_2$
- 0.1 M Copper nitrate $Ca(NO_3)_2$
- 150 ml beakers (4)
- Reaction plate, 96-well
- Toothpicks
- Distilled water

Lab Concepts

1-Limiting reagent:
The limiting reagent in a chemical reaction is a reactant that is totally consumed when the chemical reaction is completed. The amount of product formed is limited by this reagent, since the

reaction cannot continue without it. To figure out the amount of product produced, it must be determined which reactant stops the chemical reaction (the limiting reagent) and which reactant is in excess (the excess reagent). One way of finding the limiting reagent is by calculating the amount of product that can be formed by each reactant; the one that produces less product is the limiting reagent.

For the reaction between lead(II) nitrate and potassium iodide, most precipitate (product) is formed when 1 mole of lead(II) nitrate and 2 moles of potassium iodide are added together, hence the balanced equation for the reactants is

$$Pb(NO_3)_2(aq) + 2KI(aq) \rightarrow Products$$

2-Balanced equation double replacement reaction:

In a double replacement reaction, both sides of an equation maintain the same number of molecules of anion as well as cation. Earlier it was found that the mole ratio of $Pb(NO_3)_2$ and KI is 1:2. From the equation

$$Pb(NO_3)_2(aq) + 2KI(aq) \rightarrow Products$$

there is one atom of Pb on the left side, two molecules of (NO_3), two atoms of K, and two atoms of I on the left-hand side of the equation. The same number of atoms and molecules of corresponding elements/compounds should also exist on the right-hand side of the equation.

Thus, the balanced equation is

$$Pb(NO_3)_2(aq) + 2KI(aq) \rightarrow PbI_2(s) + 2KNO_3(aq) \quad \text{------------Eq(1)}$$

3-Determining the limiting reagent:

The mole ratio from Eq(1) for $Pb(NO_3)_2$ and KI is 1:2. So, if the number of moles of $Pb(NO_3)_2$ is less than half of the number of moles of KI, then $Pb(NO_3)_2$ is the limiting reagent, whereas if the number of moles of KI is less than twice of moles of $Pb(NO_3)_2$, then KI is the limiting reagent.

Procedure:

1-A)-A 0.1M solution of Copper(II) Nitrate (also called cupric nitrate) and 0.1M solution of Potassium Iodide were prepared. Copper(II) Nitrate has a molar mass of 187.6 g/mol, and Potassium Iodide has a molar mass of 166 g/mol. So, 1.87 g of Copper(II) Nitrate was added to 100 ml of distilled water into a 150 ml beaker labeled "A", and 1.7 g of Potassium Iodide was added to 100 ml of distilled water into a 150 ml beaker labeled B.

1-B)-Similarly, 0.1 M of Calcium Nitrate and 0.1M of Sodium Oxalate solution were prepared. Based on the molar mass, 1.6 g of Calcium Nitrate was added to 100 ml of distilled water into a 150 ml beaker labeled "C", and 1.3 g of Sodium Oxalate was added to 100 ml of distilled water into a 150 ml beaker labeled "D".

2-A)-The solutions in beakers "A" and "B" were stirred well. The solution, Copper(II) Nitrate, from beaker "A" was taken, and one drop into well A1, two drops into well B1, three drops into well C1, four drops into D1, and five drops into E1 were dropped. Similarly, the solution, Potassium Iodide, from beaker "B" was taken, and five drops into well A1, four drops into well B1, three drops into well C1, two drops into D1, and one drop into E1 were dropped. The data were recorded in Table-2-1

2-B)-The solutions in beakers "C" and "D" were stirred well. The solution, Calcium Nitrate, from beaker "C" was taken, and one drop into well A2, two drops into well B2, three drops into well C2, five drops into D2, and seven drops into E2 were dropped.

Similarly, the solution (Sodium Oxalate) from beaker "D" was taken, and five drops into well A2, four drops into well B2, three drops into well C2, two drops into D2, and one drop into E2 were dropped. The data were recorded in Table-2-2

3) After precipitates settled down, the observation was made for wells containing the maximum amount of precipitate and noted down in the table.

Results and Analysis

Well	A1	B1	C1	D1	E1
Drops of Copper(II) Nitrate	1	2	3	4	5
Drops of Potassium Iodide	5	4	3	2	1
Ratio of Copper(II) Nitrate to Potassium Iodide	1:5	1:2	1:1	2:1	5:1
Well with amount of precipitate		*Observed maximum amount of precipitate*			

Table-2-1: Reaction of different proportions of Copper(II) Nitrate + Potassium Iodide

The maximum amount of precipitate was observed in well B1, which contains 2 moles of copper nitrate and 4 moles of potassium iodide having a mole ratio of 1:2, as marked in Table-2-1. So, it is verified that in this chemical reaction the mole ratio of reactants cupric iodide and potassium iodide is 1:2.

Copper nitrate reacts with potassium iodide to form copper (II) iodide (also called cupric iodide) and potassium nitrate. There are one atom of Cu, two molecules of (NO_3), two atoms of K, two atoms of I on the left side of the equation, and likewise, the same number of atoms and molecules should exist on the right side of the equation. So, the balanced chemical equation is:

$$Cu(NO_3)_2 + 2KI \rightarrow CuI_2 + 2KNO_3 \quad \text{------------------Eq(2)}$$

It is also noted that in well A1, potassium iodide is limiting reagent, whereas in well C1, D1, E1 cupric nitrate is limiting reagent.

Well	A2	B2	C2	D2	E2
Drops of Calcium Nitrate	1	2	3	4	5
Drops of Sodium Oxalate	5	4	3	2	1
Ratio of Calcium Nitrate to Sodium Oxalate	1:5	1:2	1:1	2:1	5:1
Well with amount of precipitate			*Observed maximum amount of precipitate*		

Table-2-2: Reaction of different proportions of Calcium Nitrate and Sodium Oxalate

The maximum amount of precipitate was observed in well C2, which contains 3 moles of calcium nitrate and 3 moles of sodium oxalate having a mole ratio of 1:1, as marked in Table-2-2. So, it is verified that in this chemical reaction the mole ratio of reactants calcium nitrate and sodium oxalate is 1:1.

Calcium nitrate reacts with sodium oxalate to form calcium oxalate and sodium nitrate. There are one atom of Ca, two molecules of (NO_3), two atoms of Na, two atoms of C, and four atoms of O on the left side of the equation, and likewise, the same number of atoms and molecules of the elements/compounds should exist on right side of the equation.

So, the balanced chemical equation is:

$$Ca(NO_3)_2 + Na_2C_2O_4 \rightarrow CaC_2O_4 + 2NaNO_3 \text{ ---------Eq(3)}$$

It is also noted that in wells A2 and B2, calcium nitrate is limiting reagent, whereas in wells D2, E2 sodium oxalate is limiting reagent.

Analysis

The exact amount of product formation is determined to further analyze the result. There are the following three points to be used to calculate the amount of product:

i-) Drop in mL:
    We assume that 1 drop of the solution equals 0.05 mL, which is a fair approximation for our experiment.

ii-) Amount in moles from concentration of solution:
    In 1 L of solution, a 1 M solution contains 1 mole of molecule. So, in 0.05 mL (or 1 drop) of solution, a 0.1 M solution

contains
$(0.1) \times (0.05) \times (10^{-3})$ moles of molecules, or (0.005) millimole of molecules.

iii-) Ratio of reactant and product:
In this experiment, the ratio of reactant and product remains constant. So, the moles of product formed in a chemical reaction is determined from the moles of reactant used up in the reaction. This is the application of the mole ratio concept.

From the balanced equation Eq(2), the ratio of cupric nitrate (reactant) and cupric iodide (product) is 1:1, whereas the ratio of potassium iodide (reactant) and cupric iodide (product) is 2:1. In the following table, the amount of moles of cupric iodide is calculated for each well.

Well	A1	B1	C1	D1	E1
Cupric Nitrate (mL)	0.05	0.1	0.15	0.20	0.25
Millimole of Cupric Nitrate	0.005	0.01	0.015	0.02	0.025
Potassium Iodide (mL)	0.25	0.20	0.15	0.1	0.05
Millimole of Potassium Iodide	0.025	0.02	0.015	0.01	0.005
Limiting Reagent	Cupric Nitrate	None	Potassium Iodide	Potassium Iodide	Potassium Iodide
Excess Reagent	Potassium Iodide	None	Cupric Nitrate	Cupric Nitrate	Cupric Nitrate
Millimole of Cupric Iodide	0.005	**0.01**	0.0075	0.005	0.0025

Table-2-3: Determination of millimole of Cupric Iodide in each well using mole ratio

From Table-2-3, it is determined that the maximum amount of precipitate of cupric iodide is formed in well B1.

From the balanced equation Eq(3), the ratio of calcium nitrate (reactant) and calcium oxalate (product) is 1:1. Similarly, the ratio of sodium oxalate (reactant) and calcium oxalate (product) is 1:1. In the following table, the amount of moles of calcium oxalate is calculated for each well.

Well	A1	B1	C1	D1	E1
Calcium Nitrate (mL)	0.05	0.1	0.15	0.20	0.25
Millimole of Calcium Nitrate	0.005	0.01	0.015	0.02	0.025
Sodium Oxalate (mL)	0.25	0.20	0.15	0.1	0.05
Millimole of Sodium Oxalate	0.025	0.02	0.015	0.01	0.005
Limiting Reagent	Calcium Nitrate	Calcium Nitrate	None	Sodium Oxalate	Sodium Oxalate
Excess Reagent	Sodium Oxalate	Sodium Oxalate	None	Calcium Nitrate	Calcium Nitrate
Millimole of Calcium Oxalate	0.005	0.01	**0.015**	0.01	0.005

Table-2-4: Determination of millimole of Calcium Oxalate in each well using mole ratio

From Table-2-4, it is determined that the maximum amount of precipitate of Calcium Oxalate is formed in well C1.

Error-Mitigation

a- The required molar concentration of solutions should be prepared with the exact mass of chemicals.

b- The solution should be dropped exactly in the required amount inside the well.

Conclusion

A mole ratio is a conversion factor that relates the amounts in moles of any two substances in a chemical reaction. The coefficients from the balanced chemical reaction tell us the proportions of the reactants and products. We can use ratios of the coefficients to convert between amounts of reactants and products in our reaction. Mole ratios allow the comparison of the amounts of any two materials in a balanced equation. Calculations can be made to predict how much product can be obtained from a given number of moles of reactant.

In the experiment, first of all, the mole ratio of reactants was observed in Table-2-1 and Table-2-2, and then the balanced equation was determined. Besides that, the limiting reagents were also found. In Table-2-3 and Table-2-4, the actual amount of precipitate formation was determined.

It was hypothesized at the beginning of this investigation that maximum precipitation in a well shows the actual mole ratio of reactants in a chemical reaction, which turns out to be true as observed in Table-2-1 and Table-2-2, as shown in balanced chemical equations Eq(3) and Eq(4).

The real-world connection of this investigation is that generally chemical reactions follow the law of conservation of mass, and the law of constant proportions, which ensures the reactants and products maintain a defined mole ratio relationship.

3 Lab 3: Paper Chromatography

Goals

To study if substances can be separated based on differences in properties, like molecular mass.

To observe how the method of chromatography can be used to separate chemical mixtures.

Introduction

Chromatography is an analytical technique commonly used for separating a mixture of chemical substances into its individual components. In all chromatography, there is a mobile phase and a stationary phase. The stationary phase is the phase that doesn't move (generally medium) and the mobile phase is the phase that does move.

In chromatography, the mixture is dissolved in a fluid called the mobile phase, which carries it through a structure holding another material called the stationary phase. The various constituents of the mixture travel at different speeds, causing them to separate. At different points in the stationary phase, the different substances of the mixture are absorbed and stop moving with the mobile phase. This is how the results of any chromatography are gotten, from the point at which the different components of the mixture stop moving and separate from the other components.

The separation is based on differential partitioning between the mobile and stationary phases. The solutes in a mobile phase go through a stationary phase. Solutes with a greater affinity for the mobile phase spend more time in this phase than the solutes that prefer the stationary phase. As the solutes move through the stationary phase they separate. This is called chromatographic development.

In paper chromatography the mobile phase is the solvent that moves through the paper carrying different substances with it; the stationary phase is the piece of paper that is placed in the solvent. Paper chromatography is also based on differential migration and uses capillary action to move the solvent through the stationary phase. Components in a mixture are separated based on their different abilities to bind or adsorb to the stationary phase, and on their different abilities to desorb or dissolve in the mobile phase. The stationary phase refers to the solid or liquid to which components in a mixture bind or adsorb. The mobile phase refers to the liquid or gas that moves the components in a mixture over the stationary phase. The different dissolved substances in a mixture are attracted to the two phases in different proportions. This causes them to move at different rates through the paper.

Separation by chromatography produces a chromatogram. A paper chromatogram can be used to distinguish between pure and impure substances:

a pure substance produces one spot on the chromatogram

an impure substance produces two or more spots

A paper chromatogram can also be used to identify substances by comparing them with known substances. Paper chromatography is generally used to separate mixtures of soluble substances such as food colorings, inks, dyes, or plant pigments.

Lab concepts

In paper chromatography, the stationary phase is a very uniform absorbent paper, whereas the mobile phase is a suitable liquid solvent or mixture of solvents. The retardation factor is a comparison of the distance traveled by a component to the distance traveled by the solvent. For chromatography to work effectively, the components of the mobile phase are required to separate out as much as possible as they move past the stationary phase. That's why the stationary phase is often something with a large surface area, such as a sheet of filter paper, a liquid deposited on the surface of a solid, or some other highly adsorbent material.

This experiment specifically uses paper chromatography, although the principle behind this specific method is the same as in other types. A carrier liquid solvent or gas is run through a medium, which is designed to separate compounds from one another. There are several reasons for this separation, including differences in the molecular masses of different substances, chemical interactions with the medium, the process of dissolving in like substances, or the viscosity of substances in the mixture.

1-Atomic mass affects the ability of solutes to separate

The copper diffused from the bromophenol blue travels further on the chromatogram, because copper has a lighter atomic mass of 64 amu, while the atomic mass of bromophenol blue is 692 amu. So, it is inferred that the bromophenol blue could not travel that far like copper because of the higher atomic mass.

Food color consists of several dyes. Different dyes advance to different heights in the dried chromatogram. This is due to different atomic masses of dyes constituting the food color.

2-Viscosities and polarity affect chromatography:

Viscosity in centipoises at STP for water is 0.89, for isopropyl alcohol is 1.95, and for glycerin is 1500. Solvents with lower viscosity move faster through a media than solvents with a greater viscosity. Therefore, water will move faster than Isopropyl alcohol, and glycerin will be slowest due to its highest viscosity.

Polar solvents dissolve polar solutes, and non-polar solvents dissolve non-polar solutes. Isopropyl alcohol is less polar than water, so it dissolves non-polar solutes, whereas water dissolves polar solutes. Organic compounds like food color, fruit juice are more soluble in isopropyl alcohol, so the constituents of organic substances move faster and to greater distances on the chromatogram. Hence, the pattern of color on the chromatogram is different, if isopropyl alcohol is used as the solvent instead of water.

3- Separation of the mixture of copper nitrate solution and bromophenol blue

If the mixture of copper nitrate solution and same concentration solution of bromophenol blue is separated using chromatography, then a blue streak about halfway up the paper is observed. This is actually bromophenol blue, which appears on the chromatogram. But copper ions diffused from bromophenol blue remains invisible.

The visualization solution, potassium iodide (KI), is used to observe the color spot of copper ions (Cu^{2+}). So, if the chromatography paper is treated with KI by dabbing a cotton swab wetted with KI around the top of the chromatogram, a brown spot appears near the top of the chromatogram. The brown color observed for copper ion (Cu^{2+}) is produced by a chemical reaction with the potassium iodide.

4- Marking start line on chromatography paper

When drawing the start line on chromatography paper, instead of pens, pencils should be used. Ink in pens contains chemicals (dyes) that can potentially separate during the chromatography process just like chemicals do. The graphite (Carbon) in a pencil is less likely to separate during chromatography and will stay on the start line.

5-Measurement of distance traveled by solutes on the chromatogram

To calculate the absolute retention factor of a solute, it is required to measure the distance traveled by the solute on the chromatogram.

The distance traveled on the chromatogram is measured from the start line on chromatography paper to the top of the streak for every solute. The distance measured from the start line to the top is used to calculate the absolute retention factor of each solute.

The convention of measuring the distance traveled for each solute should be the same, so that the relative retention factor of each solute is consistent.

6-Calculating Retention Factor

The retention factor $R_f = \dfrac{D_{solute}}{D_{solvent}}$

Example: On a chromatogram, the distance traveled by the solvent is 7.4 cm, whereas the distance traveled by the blue dye is 7.1 cm. Find the retention factor of solvent and blue dye.

The retention factor of solvent $R_f = \dfrac{7.4}{7.4} = 1$

The retention factor of blue dye $R_f = \dfrac{7.1}{7.4} = 0.959$

Hypothesis and variables

Hypothesis: If the molecular mass of a solute is low, then it will be more mobile resulting in a greater retention factor of solute.

Independent Variable: 0.1M copper nitrate solution, bromophenol solution, potassium iodide, black felt pen, food color; beet juice, sketch pen.

Dependent Variable: Distance traveled by solutes, retention factor.

Controlled Variable: Distilled water as solvent, 0.1M concentration of copper nitrate solution, chromatography paper, the temperature of the solvent, pH of solvent.

Materials

- Beaker, glass 50 mL
- Beaker, plastic 150 mL

- Chromatography paper
- Bromophenol blue
- Copper nitrate 0.1M
- Potassium iodide 0.1M
- Reaction plate, 96-well
- Cotton Swab
- Food Coloring soliton
- Pen, wet erase black felt-tip

Procedure

1-A line 1 cm from the edge is drawn on a 10 cm by 10 cm sheet of chromatography paper.

2-Small pencil marks 2 cm apart along the line were made, starting 1 cm from the edge.

3-Three drops of 0.1M copper nitrate solution were mixed with three drops of bromophenol solution in one well of a 96-well reaction plate. After mixing the two solutions, a fine tip pipette was used to apply a drop of solution in the left-most pencil mark on chromatography paper. Two more drops of the solution were added after each drop added on paper dried completely.

4-A small toothpick was used to place a small dot of food color directly on one of the pencil spots. Two more dots of food color were added exactly on the same pencil spot, allowing the spot to dry between additions.

5-Three dots of black felt tip pen was placed on another pencil mark. The dots were placed one by one allowing the spot to dry before each addition.

6-In a similar fashion, three dots of beet juice were placed on another pencil mark.

7-Following the same procedure, three dots of blue color sketch pen were placed on another pencil mark.

8-After all the spots were dry, the paper was rolled into a cylinder and the edges were taped.

9-The cylindrical shaped paper was placed with pencil-mark down in a 150 mL plastic beaker containing distilled water to a depth of about 3 mm.

10-The paper was kept in middle without touching the edges of the beaker.

11-After 20 minutes, the paper was lifted from the beaker, the tape was removed, and the paper was left to dry.

12-Along the first pencil mark, where copper nitrate solution was supposed to travel was dabbed with potassium iodide using a cotton swab. This made the colorless copper ion on the chromatography paper visible.

13-The water had risen 103 mm on the chromatography paper.

14-The distances of color streaks above each of the spots were measured from the line drawn by pencil to the top of the color streak. The results were recorded in Table-3-1.

Results & Analysis

The water had risen 103 mm, and initially, the beaker was filled with 3 mm. Therefore, the distance traveled by the solvent front is

100 mm, or 10 cm.

The distance traveled by substances is measured from the top of the color streak to vertically down the start line (marked by pencil), which is at 1 cm from the edge.

Component	D (cm)
Solvent front	10
Copper ion	8.9
Bromophenol blue	7.8
Food color	8.1
Black felt pen	8.4
Beet juice	6.3
Sketch pen	7.6

Table-3-1 Distance traveled for different components

For each of the solutes and solvent front, the retention factor is calculated in Table-3-2.

The retention factor $R_f = \dfrac{D_{solute}}{D_{solvent}}$

Component	Retention factor
Solvent front	1
Copper	0.89
Bromophenol blue	0.78
Food color	0.81
Black felt pen	0.84
Beet juice	0.63
Sketch pen	0.76

Table-3-2 Distance traveled for different components

From Table-3-2, it is observed that copper has a higher retention factor, 0.89, than the retention factor, 0.78, of Bromophenol blue. Copper traveled further than the bromophenol blue, because it has a lighter atomic mass of 64 amu, whereas the atomic mass of bromophenol blue is 692 amu. So, it could be inferred that the bromophenol blue could not travel that far like copper because of the higher atomic mass.

Similarly, different retention factor of components implies the difference in molecular mass of components. Sometimes, the retention factor is not observed accurately, if the solute is not soluble in the solvent, but in this investigation, all components used are well soluble in water.

Core points of analysis

1-Chromatogram development dependent upon the polarity of the solvent

If the components do not get separated well on a chromatogram developed for water used as a solvent, where either a solute does not show up on a chromatogram, or has a very low retention factor, then it implies that solutes are not well dissolved into the solvent. The procedure should be modified by replacing the solvent, water, with solvents with medium polarities, such as acetone, vinegar, isopropyl alcohol, etc.

2-Appropriate solvent for different solutes

a. Newspaper inks: The carrier solution that would work best with newspaper inks is alcohol, as alcohol can dissolve inks.

b. Salts: Water solvent can be used to carry salts, as salts are soluble in water.

c. Fatty acids: Oil solvents can be used to carry fatty acids, as they're both hydrophobic substances.

3-Different retention factors of color spots on chromatogram for different dyes

The separated dye spots have different lengths, because some solutes are heterogeneous in nature having non-uniform molecular mass. If the solutes are uniformly homogenous of the same molecular mass, then the dye-spot would be centered at one place.

Another reason could be due to absorbability. The dye spots on the chromatogram show up at different lengths depending on their absorbability by the paper; the substances become less visible as they travel further along the paper and get absorbed in the process.

4-Relationship of molar mass and distance traveled on the chromatogram

The molar mass of

FD&C Yellow#5 is 534 g/mol;

FD&C Red#40 is 496 g/mol;

FD&C Blue#1 is 793 g/mol;

Based on the molecular mass, the solutes would travel different distances on the chromatogram. Due to the lightest molecular mass of FD&C Red, it will travel the greatest distance and will be on top. Similarly, FD&C Yellow will be in middle, and FD&C Blue being the heaviest will be at the lowest position on the chromatogram.

5-Distance traveled by a solute and its retention factor

Distance traveled by a solute on the chromatogram is directly proportional to its retention factor. For three samples having solutes of retention factors 0.33, 0.50, and 0.72, the corresponding distances traveled on a chromatogram is visualized in Figure-3-3.

Chromatogram sketch for given retention factors of the samples:

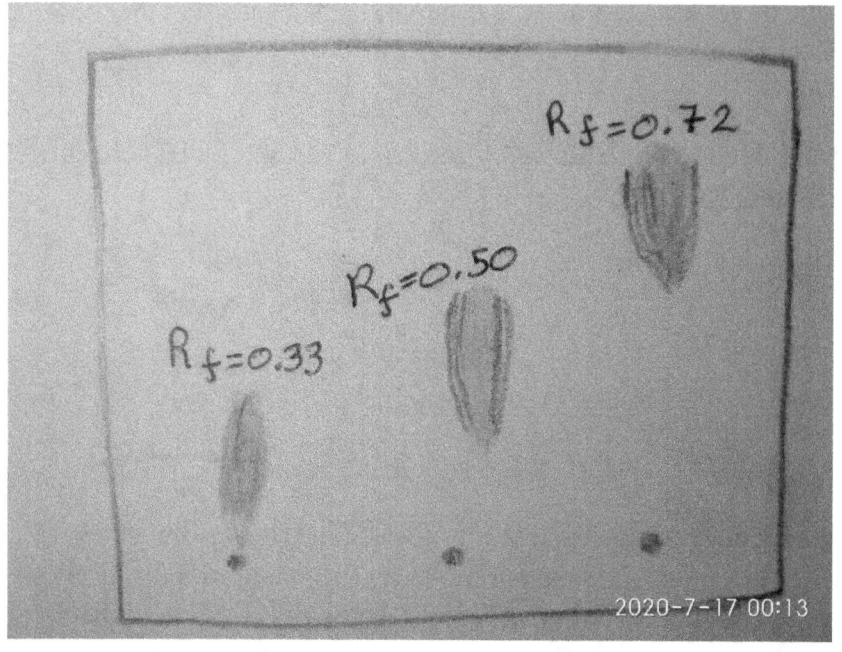

Figure-3-3 Retention factor and distance traveled for different solutes

Error-Mitigation

a- The chromatography paper strip is dipped in the solvent in such a way that the spot of the mixture is above the solvent level and the movement of the solvent front is not zig-zag.

b- It is ensured that the filter paper strip hangs freely in the jar.

c- Mixture is placed as small and circular dots on the chromatography paper.

d- The chromatography paper strip is made perfectly dry before developing the spots.

e-Once the experiment is set, the paper and beaker are not disturbed as long as the chromatogram development is in process.

Conclusion

Chromatography is a process used to separate the components of a mixture. A mixture is injected into a chromatography column, where it lands on a substrate, also known as the stationary phase. The stationary phase may be polar, attracting polar substances, or nonpolar, attracting nonpolar substances.

It was hypothesized at the beginning of this investigation that if the molecular mass of a solute is low, then it will be more mobile resulting in a greater retention factor of solute, which turns out to be true. As observed in Table-3-2, copper has a higher retention factor of 0.89 than the retention factor 0.78 of Bromophenol blue. Copper traveled further than the bromophenol blue, because it has a lighter atomic mass of 64 amu, whereas the atomic mass of bromophenol blue is 692 amu.

The real-world connection of chromatography is its wide application in different fields. Chromatography is now widely used in forensic science, in pollution monitoring, and for studying complex mixtures in such things as food, perfume, petrochemical, and pharmaceutical production. The air samples are examined using chromatography for identifying small concentrations of unknown pollutants in air and water samples. One of chromatography's big advantages is that it works with low concentration samples.

Lab 4: Types of Chemical Reactions

Goals

To study the transitions that occur with chemical changes.
To study the pattern and rules of solubility
To use empirical evidence to determine the type of chemical changes.
To study ways of writing chemical changes as a balanced reaction.

Introduction

Chemical reactions are the process where the chemical composition of the reactants is changed into products. There are several changes during a chemical reaction, which include a spontaneous color change, precipitation of a solid, a temperature change, change in pH, or bubbling.

For reactions in an aqueous solution, a double displacement reaction could also be a neutralization reaction. During a redox reaction, the oxidation number of one or more elements is changed in the process of the chemical reaction. These reactions can also be classified as synthesis, single replacement, or double replacement type of reactions depending on the reactants and products involved.

A double replacement reaction involves the exchange of two substances and is often characterized by precipitation. A precipitation reaction is one in which dissolved substances react to form one or more solid products. These reactions generally involve the exchange of ions between ionic compounds in an aqueous solution.

The extent to which a substance may be dissolved in a solvent is quantitatively expressed as its solubility, which is the maximum concentration of a substance that can be achieved under specified conditions. Substances with relatively large solubility are said to be soluble. A substance precipitates, when its concentration in solution exceeds its solubility. Substances with relatively low solubility are said to be insoluble, and these are the substances that readily precipitate from the solution.

Example of double replacement reaction:

$$2AgNO_3\ (aq) + CaCl_2(aq) \rightarrow Ca(NO_3)_2\ (aq) + 2AgCl\ (s)$$

Lab concepts

Solubility rules for aqueous solution

Ions That Form Soluble Compounds	Exceptions	Ions That Form Insoluble Compounds*	Exceptions
Group 1 ions (Li^+, Na^+, etc.)		carbonate (CO_3^{2-})	when combined with Group 1 ions or ammonium (NH_4^+)
ammonium (NH_4^+)		chromate (CrO_4^{2-})	when combined with Group 1 ions, Ca^{2+}, Mg^{2+}, or ammonium (NH_4^+)
nitrate (NO_3^-)			
acetate ($C_2H_3O_2^-$ or CH_3COO^-)		phosphate (PO_4^{3-})	when combined with Group 1 ions or ammonium (NH_4^+)
hydrogen carbonate (HCO_3^-)		sulfide (S^{2-})	when combined with Group 1 ions or ammonium (NH_4^+)
chlorate (ClO_3^-)		hydroxide (OH^-)	when combined with Group 1 ions, Ca^{2+}, Ba^{2+}, Sr^{2+}, or ammonium (NH_4^+)
halides (Cl^-, Br^-, I^-)	when combined with Ag^+, Pb^{2+}, or Hg_2^{2+}		
sulfates (SO_4^{2-})	when combined with Ag^+, Ca^{2+}, Sr^{2+}, Ba^{2+}, or Pb^{2+}	*compounds having very low solubility in H_2O	

Table-4-1: Solubility rules

Periodic Table

Group 1	2											13	14	15	16	17	18
1 H		Group names*															2 He
		1 Hydrogen & the alkali metals		14 Carbon Group													
		2 Alkaline earth metals		15 Pnictogens													
3 Li	4 Be	11 Coinage metals (Cu, Ag & Au)		16 Chalcogens								5 B	6 C	7 N	8 O	9 F	10 Ne
		12 Volatile metals		17 Halogens													
		13 Boron Group		18 Noble gases													
11 Na	12 Mg	3	4	5	6	7	8	9	10	11	12	13 Al	14 Si	15 P	16 S	17 Cl	18 Ar
19 K	20 Ca	21 Sc	22 Ti	23 V	24 Cr	25 Mn	26 Fe	27 Co	28 Ni	29 Cu	30 Zn	31 Ga	32 Ge	33 As	34 Se	35 Br	36 Kr
37 Rb	38 Sr	39 Y	40 Zr	41 Nb	42 Mo	43 Tc	44 Ru	45 Rh	46 Pd	47 Ag	48 Cd	49 In	50 Sn	51 Sb	52 Te	53 I	54 Xe
55 Cs	56† Ba	71 Lu	72 Hf	73 Ta	74 W	75 Re	76 Os	77 Ir	78 Pt	79 Au	80 Hg	81 Tl	82 Pb	83 Bi	84 Po	85 At	86 Rn
87 Fr	88‡ Ra	103 Lr	104 Rf	105 Db	106 Sg	107 Bh	108 Hs	109 Mt	110 Ds	111 Rg	112 Cn	113 Uut	114 Fl	115 Uup	116 Lv	117 Uus	118 Uuo

Table-4-2: Periodic table

Prediction of precipitate reaction from Table-4-1:

General rules for the formation of precipitates can be used to predict a precipitate reaction.

The solubility of ionic compounds in water at STP from Table-4-1 is summarized as:

1. All compounds of the ammonium ion (NH_4) are soluble.
2. All compounds of Alkali metal (Group 1, or, Group IA) cations, are soluble.
3. All nitrates are soluble.
4. All acetates (ethanoates) are soluble.
5. All chlorides, bromides, and iodides are soluble except silver, lead, and mercury(I)
6. All sulfates are soluble except silver, lead, mercury(I), barium, strontium, and calcium
7. All carbonates, sulfites, and phosphates are insoluble except ammonium (NH_4) and alkali metal (Group 1, or, Group IA) cations.
8. All hydroxides are insoluble except ammonium (NH_4), barium and alkali metal (Group 1, or, Group IA) cations.
9. All sulfides are insoluble except ammonium, alkali metal (Group 1, or, Group IA) cations and alkali earth metal (Group 2, or, Group IIA) cations.
10. All oxides are insoluble except calcium, barium, and alkali metal (Group 1, or, Group IA) cations; these soluble substances react with the water (hydrolyze) to form hydroxides, which is known as a hydrolysis reaction.

Example: Aqueous solution of silver nitrate, $AgNO_3$(aq), is added with dilute hydrochloric acid, HCl (aq).

Step-1: All possible anions and cations

		\multicolumn{2}{c}{**ions from $AgNO_{3(aq)}$**}
		silver ion $Ag^+_{(aq)}$	nitrate ion $NO_3^-{}_{(aq)}$
ions from $HCl_{(aq)}$	hydrogen ion $H^+_{(aq)}$		
	chloride ion $Cl^-_{(aq)}$		

Step-2: All possible ionic compounds that can be produced

		ions from $AgNO_{3(aq)}$	
		silver ion $Ag^+_{(aq)}$	nitrate ion $NO_3^-{}_{(aq)}$
ions from $HCl_{(aq)}$	hydrogen ion $H^+_{(aq)}$	-	hydrogen nitrate (nitric acid) HNO_3
	chloride ion $Cl^-_{(aq)}$	silver chloride AgCl	-

Step-3: Applying the solubility rules from Table-4-1,

(i) All nitrates are soluble, so hydrogen nitrate (nitric acid) is soluble and will not form a precipitate, $HNO_3(aq)$.

(ii) All chlorides are soluble except silver, lead, and mercury(I), so silver chloride is insoluble and will form a precipitate, $AgCl(s)$.

Solubility rule and chemical equations:

A mixture of silver nitrate solution ($AgNO_3$) and calcium chloride solution ($CaCl_2$) are mixed.

The precipitate can be determined by exchanging the two ions and applying the solubility rules from Table-4-1. As all nitrates are soluble, so calcium nitrate will be soluble. But chloride anion combined with silver cation is insoluble and a precipitate of silver chloride, an ionic compound, will be formed.

Hence, the molecular and ionic reactions are expressed as:

Molecular equation:

$$2AgNO_3\ (aq) + CaCl_2(aq) \rightarrow Ca(NO_3)_2\ (aq) + 2AgCl\ (s)$$

Ionic Equation:

$$2Ag^+(aq) + 2NO_3^-\ (aq) + Ca^{2+}\ (aq) + 2Cl^-\ (aq) \rightarrow 2AgCl\ (s) + Ca^{2+}\ (aq) + 2NO_3^-(aq)$$

The net ionic equation:

$$Ag^+\ (aq) + Cl^-\ (aq) \rightarrow AgCl\ (s)$$

Difference between molecular and ionic compounds:

Molecular compounds are substances formed when atoms are linked together by sharing of electrons through covalent bonds.

In ionic compounds, electrons are completely transferred from one atom to another so that a cation (positively charged ion) and an anion (negatively charged ion) form. The strong electrostatic attraction between adjacent cations and anions is known as an ionic bond.

- Molecular compounds are formed between two non-metals, whereas ionic compounds are formed between metals and non-metals.
- Molecular compounds can be in any physical state - solid, liquid, or gas, whereas ionic compounds are always solid and crystalline in appearance.
- Molecular compounds are poor electrical conductors while ionic compounds are good conductors.

Some of the double replacement reactions form molecular compounds instead of ionic compounds. Any sulfate solution added with hydrochloric acid produces the molecular compounds, sulfur dioxide gas, and water.

$$2HCl\ (aq) + NO_2SO_3(aq) \rightarrow SO_2(g) + H_2O\ (l) + 2NaCl\ (aq)$$

Hypothesis and variables

Hypothesis: If carbonate anion pairs up with cation of non-alkali metal in a double displacement reaction, then precipitation takes place.

Independent Variable: Given reagents to test solubility.

Dependent Variable: Change in color, precipitation, and other physical changes.

Controlled Variable: Concentration of reactants, temperature, and pH of reactants.

Materials

- Reagents:
    - 0.1 M acetic acid
    - 0.1 M calcium nitrate
    - 0.1 M copper (II) nitrate
    - 0.1 M nickel (II) nitrate
    - 0.1 M zinc nitrate
    - 0.1 M potassium iodide
    - 0.1 M potassium hydroxide
    - 0.1 M sodium oxalate
    - 0.1 M sodium sulfate
- Reaction plate 24-well
- Reaction plate 96-well
- 0-6 pH paper
- Digital balance
- Beaker, plastic 150 mL
- Weighing boat (2 oz plastic cup)
- Toothpicks

Procedure

Part I

1- Three drops of 0.1 M calcium nitrate in wells A1 through A4, three drops of 0.1 M copper (II) nitrate in wells B1 through B4, three drops of 0.1 M nickel (II) nitrate in wells C1 through C4, and three drops of 0.1 M zinc nitrate in wells D1 through D4 were put.

2- Three drops of 0.1M potassium iodide in wells A1 through D1 (the first column), three drops of 0.1M potassium hydroxide in wells A2-D2, three drops of 0.1M sodium oxalate in well A3-D3, and three drops of 0.1M sodium sulfate in A4-D4 were put.

3- A chemical reaction was indicated by the change in color or precipitate formation. The changes in reaction wells were recorded in Table-4-3. For no noticeable change "NR" was recorded.

Part II

1- A sheet of ordinary white paper, knowing that it contains starch, was taken.

2- A drop of potassium iodide was added to paper right next to the copper nitrate, without solutions touching each other.

3- The changes in the color of the chemical reaction were recorded in Table-4-4 that takes place with individual solutions and the starch of the paper.

4- The toothpick was used to mix the two solutions together on the paper and any change in color was recorded in Table-4-4.

5- The paper containing the solutions was discarded after the experiment.

Part III

1- A solution of 0.1 M $NaHCO_3$ was prepared and then reacted with HCl solution. To prepare a 0.1 M solution of $NaHCO_3$, steps 2 and 3 were performed.

2- 0.84 g of baking soda ($NaHCO_3$) is weighed in the tared weighing boat (2 oz plastic cup).

3- The 150 mL plastic beaker is filled to the 30 mL mark with distilled water. The 0.84 g of baking soda was poured into the water in the beaker and stirred with the glass stir rod to dissolve. Thereafter, water was added to the 100 mL mark, stirred again to mix properly.

4- Using the dropping bottle, 5 drops of 0.1 M HCl were added to a well in the 24 well reaction plate.

5- Using the graduated pipet, 5 drops of 0.1 M $NaHCO_3$ were added to the HCl solution. The reaction was observed and recorded in Table-4-5.

6- Extra drops of $NaHCO_3$ were added to the well until there was no more reaction, and the observations were recorded in Table-4-5.

7- The sodium hydrogen carbonate solution was discarded, and the 24 well reaction plate was rinsed.

Results

The change in reaction wells is recorded in Table-4-3.

		1	2	3	4
		KI	KOH	$Na_2C_2O_4$	Na_2SO_4
A	$Ca(NO_3)_2$	NR	Precipitate	Precipitate	Milky white
B	$Cu(NO_3)_2$	Redox; Yellowish-Brown	Greenish blue	Light blue	NR
C	$Ni(NO_3)_2$	NR	Green	Light green	NR
D	$Zn(NO_3)_2$	NR	White	White	NR

Table-4-3: Change in reaction wells

Table-4-4 shows the data for observations of part II of the procedure.

		Starch	Starch + Copper (II) nitrate
Potassium Iodide	Orange	Deep blue	Deep blue
Copper (II) nitrate	Blue	NR	NR

Table-4-4: Iodine starch test

Table-4-5 shows the data for observations of part III of the procedure.

	0.1 M HCl
5 drops of 0.1 M $NaHCO_3$	colorless, Bubbles
5 more drops of 0.1 M $NaHCO_3$	colorless

Table-4-5: Chemical reaction between $NaHCO_3$ and HCl

Analysis / Discussion

Part 1 of the procedure is carried out to investigate the validity of solubility rules. There were various combinations of nitrate anions with alkali metal anions used in double replacement reactions. The changes in reaction wells were observed in Image-4-6 for part-1 of the procedure.

Part II of the procedure is performed to investigate the reduction of copper (II) to copper (I).

Image-4-6: Changes in reaction wells for part-1 of the procedure

In part III the procedure, reaction of sodium bicarbonate and HCl produces a salt and carbonic acid, which readily decomposes to carbon dioxide and water:

$$NaHCO_3 + HCl \rightarrow NaCl + H_2CO_3$$
$$H_2CO_3 \rightarrow H_2O + CO_2\,(g)$$

Core points of analysis

1-Examples of double replacement reactions and formation of precipitates

1-a- Aqueous solution of calcium chloride and sodium carbonate

The calcium cations pair up with the carbonate anion to form the insoluble calcium carbonate, $CaCO_3$. This agrees with solubility rules.

Molecular equation:

$$Na_2CO_3\ (aq) + CaCl_2 \rightarrow CaCO_3\ (s) + 2NaCl\ (aq) \quad \text{--------Eq(1)}$$

Ionic Equation:

$$2Na^+(aq) + CO_3^{2-}\ (aq) + Ca^{2+}\ (aq) + 2Cl^-\ (aq) \rightarrow CaCO_3\ (s) + 2Na^+\ (aq) + 2Cl^-\ (aq)$$

The net ionic equation:

$$CO_3^{2-}\ (aq) + Ca^{2+}\ (aq) \rightarrow CaCO_3\ (s)$$

1-b- Aqueous solution of Acetic acid and potassium hydroxide

The acetate anion pairs up with potassium cation to form soluble potassium acetate. This is in accordance with the solubility rule.

Molecular equation:

$HC_2H_3O_2\,(aq) + KOH\,(aq) \rightarrow KC_2H_3O_2(aq) + H_2O(l)$ --------Eq(2)

Ionic equation:

$HC_2H_3O_2\,(aq) + K^+\,(aq) + OH^-\,(aq) \rightarrow$

$\qquad K^+\,(aq) + C_2H_3O_2^-\,(aq) + H_2O\,(l)$

Net ionic equation:

$HC_2H_3O_2\,(aq) + OH^-\,(aq) \rightarrow\; + C_2H_3O_2^-\,(aq) + H_2O\,(l)$

1-c- Aqueous solution of silver nitrate and potassium phosphate

Silver cation pairs up with phosphate anion to form silver phosphate, which is insoluble. According to solubility rules, all phosphates are insoluble, except Na^+, K^+, ammonium. When silver nitrate is mixed with something that is only soluble because of the potassium ion, then an insoluble (solid) silver phosphate is formed.

Molecular equation:

$3AgNO_3\,(aq) + K_3PO_4\,(aq) \rightarrow Ag_3PO_4\,(s) + 3KNO_3\,(l)$ ------Eq(3)

Ionic equation:

$3Ag^+ (aq) + PO_4^{3-} (aq) + 3K^+ (aq) + 3NO_3^- \rightarrow$
$Ag_3PO_4 (s) + 3K^+ (aq) + 3NO_3^-$

Net ionic equation:

$3Ag^+ (aq) + PO_4^{3-} (aq) \rightarrow Ag_3PO_4 (s)$

1-d-Aqueous solutions of carbonic acid and potassium hydroxide

The potassium cation pairs up with carbonate anion and forms soluble potassium carbonate. This agrees with the solubility rule.

Molecular equation:

2KOH (aq) + H2CO3 (aq) → K2CO3 (aq) + 2H2O (l) ----------Eq(4)

Ionic Equation:

$H_2CO_3 (aq) + K^+(aq) + 2OH^-(aq) \rightarrow$
$\qquad K^+(aq) + 2H_2O (l) + CO_3^{2-}(aq)$

Net Ionic Equation:

$H_2CO_3 (aq) + 2OH^-(aq) \rightarrow 2H_2O (l) + CO_3^{2-}(aq)$

2-Chemical equation and formation of the molecular compound in part III of the procedure

In part III of the procedure, carbonic acid, H_2CO_3, was produced and decomposed into carbon dioxide gas and water.

The molecular equation:

$$H_2CO_3 \text{ (aq)} \rightarrow H_2O \text{ (l)} + CO_2 \text{ (g)}$$

The ionic equation:
$$H^+(aq) + HCO_3^- (aq) \rightarrow H_2O \text{ (l)} + CO_2 \text{ (g)}$$

The decomposition of carbonic acid appeared in the form of bubbles due to the formation of carbon dioxide gas, which is a molecular compound having covalent bonds.

3- Oxidation.-reduction of copper (II) to copper (I)

$$2Cu^{2+} + 2I \rightarrow 2Cu^+ + I_2$$

Reduction: Copper reduces from copper (II) to copper (I) by gaining electrons

$$2Cu^{2+} + 2e^- \rightarrow 2Cu^+$$

Oxidation: Iodine turns into iodide losing electrons

$$2I \rightarrow I_2 + 2e^-$$

4- Illustration of oxidation-reduction reaction of copper(II) iodide

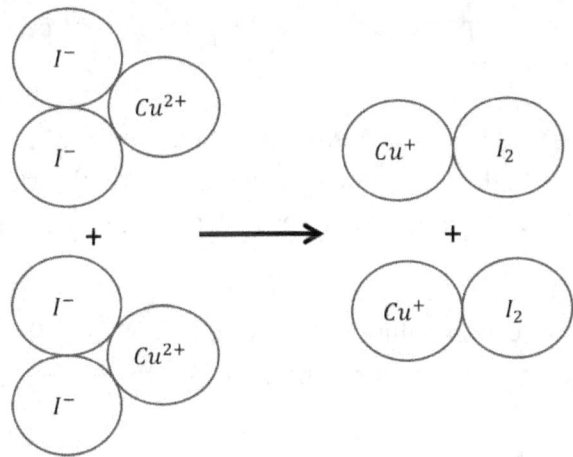

Figure-4-8 Oxidation-reduction of copper(II)iodide

5-Illustration of ionic reaction between calcium nitrate and sodium oxalate

$Ca(NO_3)_2 + (Na_2C_2O_4) \rightarrow CaC_2O_4 + 2NaNO_3$

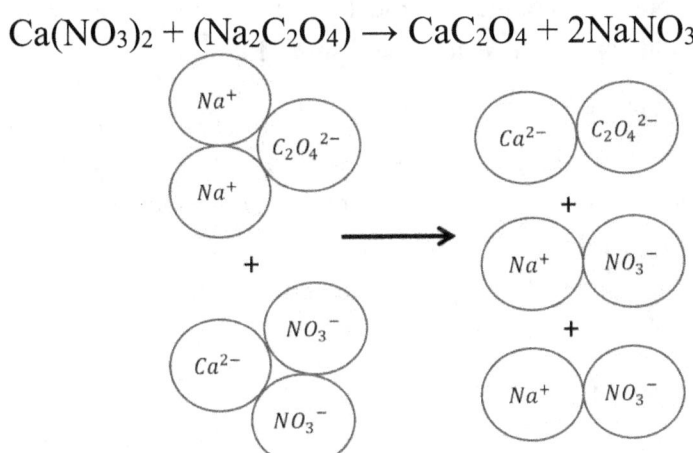

Figure-4-7 Ions before and after the reaction

Risk-Error-Mitigation

a-Pressure and temperature affect solubility, so controlled variables must be kept constant.

b-The solutions used in the experiment may have been of slightly different concentration than assumed. Several trials of the experiment will provide better results, and mitigate the effect of slight differences.

c-Some of the cation solutions are toxic, so must be handled with care. Safety measures must be observed strictly as directed.

Conclusion

In the investigation, the precipitate formation was observed in a chemical reaction using solubility rules. This lab also included forming molecular and ionic chemical equations after observing experimental data and using established solubility rules.

A lot of ionic compounds dissolve in water, dissociating into individual ions, but when two ions find each other that form an insoluble compound, they suddenly isolate themselves from the solution called a precipitation reaction. The net ionic equation shows only the chemical species that are involved in a reaction, while the complete ionic equation also includes spectator ions.

It was hypothesized at the beginning of this investigation that if carbonate anion pairs up with cation of non-alkali metal in a double displacement reaction, then precipitation takes place, which turns out to be true as mentioned in Eq(1). Calcium is a non-alkali metal cation that pairs up with the carbonate anion to form the insoluble calcium carbonate, $CaCO_3$.

There is a real-world connection, as the types of reactions investigated are common in nature. Redox reactions are used widely in industry for the production of several commodities and specialty chemicals. Precipitation reactions also play a central role in many chemical analysis techniques, including spot tests used to identify metal ions and gravimetric methods for determining the composition of matter.

Lab 5: Colorimetry

Goals
To study the relationship of transmittance, concentration, and light path length.
To determine the concentration of a colored solution by using the absorbance of light.

Introduction

The particle nature of light is most often used to explain the absorption and emission properties of atoms and molecules. Molecules absorb or emit energy only in discrete amounts or packets called quanta.

It is not so easy to measure the actual absorbance of light. Instead, we measure transmittance or the fraction of light that is able to pass through a solution of molecules

There are three things that affect the amount of light emerging from the sample.

1. The concentration of molecules in the solution affects the transmittance. Each molecule can absorb light. As the number of molecules in the solution is increased, the photons absorbed is also increased. Therefore, as the concentration of a sample is increased, the transmittance is correspondingly decreased.
2. The transmittance is affected by specific properties of the molecules. Molecules absorb light at different efficiencies and energies. Therefore, the transmittance is dependent upon the specific molecule in the solution and the wavelength of light being passed through the sample.
3. The length of the sample path affects the transmittance. By increasing the pathway that the light must travel through your sample, you are increasing the number of molecules

that will interact with the light; in effect, you are increasing the apparent concentration.

One way to measure the concentration of solutes in solution is colorimetry. Certain chemicals absorb certain wavelengths of light in solution and reflect certain wavelengths of light. For example, blue solutions absorb all colors of light except blue, which is why the solution appears blue.

The absorbance has a logarithmic relationship to the transmittance; with an absorbance of value 0 corresponding to a transmittance of 100%, and an absorbance of value 1 corresponding to 10% transmittance. Absorbance is quantified by Lambert-Beer Law as the quantity of light that is absorbed (and is not transmitted) by the solution.

$$A = 2 - \log(\%T) \quad \text{-----Eq(1)}$$

where **A** is the absorbance, and **%T** is the percentage transmittance.

The absorbance for a specific concentration of a solution with a fixed path length varies directly with the absorptivity coefficient of the solution. This relationship is known as Beer's law.

$$A = a\,b\,c \quad \text{-----Eq(2)}$$

where **A** is absorbance, **a** is the molar absorptivity coefficient, **b** is the path length in cm, corresponding to the distance light travels through the solution, and **c** is the concentration of the solution.

Hypothesis and variables

Hypothesis: If the intensity of the color of two solutions is the same, then the product of path length and concentration of the solutions are the same.

Independent Variable: Intensity of color, stock solution, water, sample solutions (A, B, C, D).

Dependent Variable: path length, inverse path length, concentration.

Controlled Variable: Light source.

Materials

- Food Dye
- Pepsi Blue
- Bug Juice
- Water
- Reaction plate, 24 well
- 2 Test tubes, 12x75 mm
- Digital balance
- Glass beaker, 50 mL
- Graduated pipet
- Plastic beaker, 150 mL
- 5 Plastic cups, 3 oz
- Notecards, 3x5 inch
- Paper
- Tape

Chemistry Lab Investigations — Saurya Singh

Lab concept

In a reaction plate, 5, 10, 15, 20 drops of the same concentration colored solution were added into four wells. Looking at the reaction plate holding it 15 cm above a white sheet of paper, the color intensity appeared different, although the concentration of the solution is the same in all the four wells.

The color intensity was strongest in 20 drops of well, followed by 15, then 10, and at last 5. This is because the light absorbance is directly proportional to the amount of color pigments in the well, and the well with 20 drops of the solution has the highest amount of color pigment, followed by 15, 10, and 5 drops.

Procedure

1-150 mL of water was put in a 150 mL plastic beaker. Three drops of food color were added to the water and stirred well. This way, the stock solution was prepared.

2-Four cups were labeled as "A", "B", "C", and "D".

3-Cup "A" was put on the digital balance and the balance was tared. 20 grams of blue dye solution was poured into cup "A" from 150 mL of stock solution. The mas of initial stock solution was recorded in Table-5-2. 10 g of water was added to the solution to get 30 g of diluted solution. The final diluted mass was recorded.

4-The step 3 was repeated for remaining cups "B", "C", and "D" preparing solutions of blue dye solution and water as shown in Table-5-1.

5-A sheet of white paper was put on a flat surface having a white light shining uniformly placed under a lamp.

6-A 3x5 inch card was taken and cut to a height of 6.5 cm to get a dimension of 6.5 cm x 12.7 cm. The card was cut in half to have two smaller cards of the same dimension 6.5 cm x 6.35 cm.

7-One of the card rectangles was rolled around the 12x75 mm test tube so that the top 10 mm of the test tube is unwrapped. The seam of the rolled card was taped so that it kept its tubular shape when the test tube was removed. This way, a rolled card tube was made. Similarly, a second rolled card tube was made.

8-Two tubes were taped together parallel to one another. Another piece of transparent tape was used to prevent the test tubes from sliding out of the rolled card tubes.

9-The 12x75 mm test tubes were rinsed and cleaned from inside and outside.

10-A pipet was used to fill one of the test tubes with the stock solution to a depth of exactly 2.0 cm from the bottom of the test tube to the bottom of the meniscus of the solution. This depth of 2 cm was treated as the light path length "b" for the stock solution.

11-The test tube with 2 cm of stock solution was placed in one of the rolled card tubes with the bottom resting on the tape.

12-The clean and empty second test tube was slid into the rolled card tube resting on the tape.

13-The two test tubes slid into the rolled card tube were held so that lit paper was viewed through the top of both test tubes.

14-Using the graduated pipet, solution A was added slowly to the empty test tube until it had the same degree of transparency and color as the 2.0 cm of the stock solution in the first test tube. It was ensured that the two test tubes had the same intensity of color.

15-The test tube with solution A was removed from the paper tube.

The height of solution A in the test tube was measured and recorded in Table-5-3.

16- The volume of solution A was calculated. The solutions were so dilute that the densities were very close to water approximated as 1.00 g per 1.00 mL. The 50 mL glass beaker was placed on the digital balance and balance was tared. The test tube with solution A was put in a glass beaker and the mass was recorded in Table-5-4. The test tube was rinsed and dried. Mass of the empty test tube was also recorded in Table-5-4.

17- The mass of the test tube was subtracted to find the mass of the solution and data was recorded in Table-5-4.

18- The steps from 12 to 17 were repeated using solution B through D.

19- As data were recorded for solution A, similarly, the data were recorded in Table-5-3 and Table-5-4 for solution B, C, and D.

Cup	Target mass of stock blue dye(g)	Target mass of water added (g)
A	20	10
B	20	20
C	20	30
D	20	46

Table-5-1 Solution preparation using stock solution and water

Results

The mass of stock dye and diluted solutions are recorded in Table-5-2.

Cup	Target mass of stock blue dye (g)	Actual mass of stock blue dye (g)	Target mass of water added (g)	Actual final diluted solution mass (g)
A	20	20.01	10	30.03
B	20	20.05	20	40.06
C	20	20.07	30	50.05
D	20	20.03	46	66.07

Table-5-2 Actual mass of solution prepared using stock solution and water

For sample solutions, A, B, C, D, the light path length corresponding to the same absorbance as the stock solution are recorded in Table-5-3.

Solution	Light path length "b" (cm)
Stock Solution	2.00
A	2.9
B	3.9
C	5.1
D	6.2

Table-5-3 Light path length for different solutions

The mass of solutions A, B, C, D is recorded in Table-5-4.

Mass of empty test tube: 0.35 g

Solution	Mass of test tube and solution (g)	Mass Solution (g)
Stock Solution	1.68	1.33
A	2.28	1.93
B	2.95	2.6
C	3.75	3.4
D	4.48	4.13

Table-5-4 Mass of test tube and solution

Analysis/Discussion

The Beer-Lambert law states that the quantity of light absorbed by a substance dissolved in a fully transmitting solvent is directly proportional to the concentration of the substance and the path length of the light through the solution.

Beer-Lambert Law also relates the transmittance of light to absorbance by taking the negative logarithmic function, base 10, of the transmittance observed by a sample, which results in a linear relationship to the intensity of the absorbing species and the distance traveled by the light.

In the investigation, the path length and concentration of four sample solutions were determined (Table-5-6) by making their absorbance the same as the stock solution, implying that the absorbance is kept constant. Thus, in Eq(2), absorbance **A** and absorptivity coefficient **a** are constant. This implies that the product of path length **b**, and concentration **c** is a constant. In Table-5-6, it is also found that the product of concentration **c**, and light path length **b** of all the four sample solutions and the stock solution are constant. Hence, the observation of this investigation is valid, as it follows Beer-Lambert's law.

In Figure-5-8, the graph is almost rectilinear, so the function is formulated based on the equation of a line on the Cartesian coordinate of X-Y axis as

$y = mx + b$, where m is the slope and b is the y-intercept.

In the experiment, the equation of inverse light path length as a function of concentration is formulated as

$I = 0.05 \, C - 0.0015$

Alternatively, the equation could also be formulated in terms of light path length "b" and concentration C as

$$\frac{1}{b} = 0.05\,C - 0.0015$$

Core points of analysis

<u>1-Determine the relative concentrations of sample solutions, assuming the concentration of stock solution to be 10 μM.</u>

Mass in grams and volume in mL are approximately the same, i.e. 1g = 1 mL, as mentioned in Table-5-5.

Assuming the concentration of stock solution be 10 μM, the relative concentrations of sample solutions are calculated using

$(C_{initial\ stock\ solution}) \times (V_{initial\ stock\ solution}) = (C_{final\ diluted\ solution}) \times (V_{final\ diluted\ solution})$

Using this formula, the concentration of sample solutions A, B, C, D are calculated and recorded in Table-5-5.

The concentration of diluted solution A for 1.93 mL

$$C_A = \frac{(1.33) \times (10)}{1.93} = 6.89\ \mu M$$

The concentration of diluted solution B for 2.6 mL

$$C_B = \frac{(1.33) \times (10)}{2.6} = 5.11\ \mu M$$

The concentration of diluted solution C for 3.4 mL

$$C_C = \frac{(1.33) \times (10)}{3.4} = 3.91 \; \mu M$$

The concentration of diluted solution D for 4.13 mL

$$C_D = \frac{(1.33) \times (10)}{4.13} = 3.22 \; \mu M$$

Solution	Volume of solution (mL)	Concentration (μM)
Stock Solution	1.33	10
A	1.93	6.89
B	2.6	5.11
C	3.4	3.91
D	4.13	3.22

Table-5-5 Volumes and concentrations of solution

2-Determine the line of best fit for the length of the light path for identical transparency as a function of the concentration of solutions.

2-A: In the Table-5-6, the concentration data from Table-5-5, the light path "b" data from Table-5-3 are recorded, and the inverse of the light path lengths are calculated.

Solution	Concentration (μM)	Light path length "b" (cm)	Inverse light path length "1/b" (cm^{-1})
Stock Solution	10	2.00	0.5
A	6.89	2.9	0.34
B	5.11	3.9	0.25
C	3.91	5.1	0.19
D	3.22	6.2	0.16

Table-5-6 Concentration, light path length, and inverse light path length of solutions

It is to be noted that the number of molecules of color pigment is directly proportional to absorbance. So, the number of molecules in sample solution A (with the path length of 2.9 cm and concentration of 6.89 μM) is the same as in the stock solution (with the path length of 2.0 cm with a concentration of 10 μM).

Similarly, the number of molecules in sample solution A (with the path length of 2.9 cm and concentration of 6.89 μM) is the same as in sample solution B (with the path length of 3.9 cm with a concentration of 5.11 μM).

2-B: Graph of light path length on the y-axis, and concentration on the x-axis is shown in Figure-5-7.

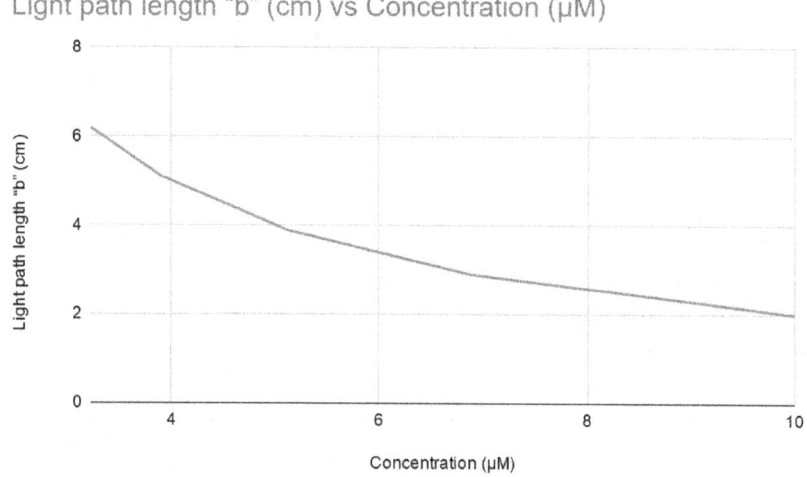

Figure-5-7 Rate of change of light path length with respect to concentration

Graph of inverse light path length on the y-axis, and concentration on the x-axis is shown in Figure-5-8.

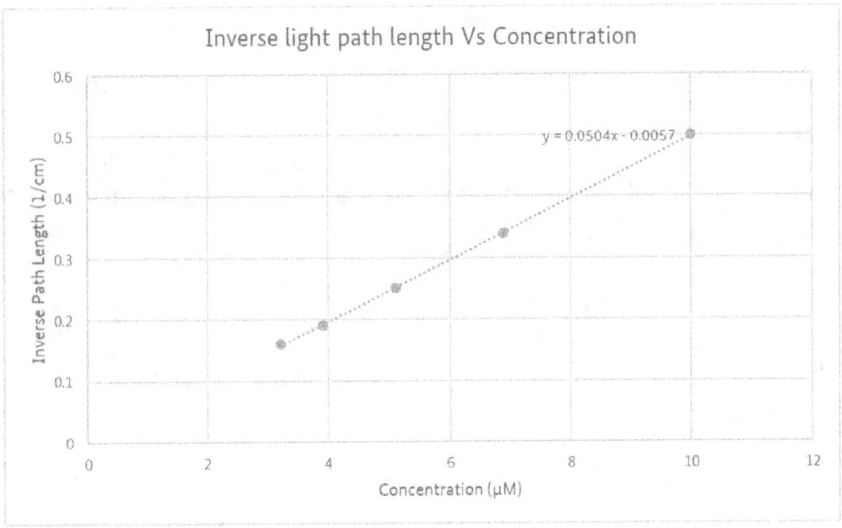

Figure-5-8 Rate of change of inverse light path length with respect to concentration

2-C: In Figure-5-8, the graph is almost rectilinear, so the function is formulated based on the equation of a line on the Cartesian coordinate of X-Y axis as

$$y = mx + b$$

where m is the slope and b is the y-intercept.

In Figure-5-8, the concentration C represents the x-axis, and inverse light length I represents the y-axis. The slope is calculated as

$$m = \frac{\Delta I}{\Delta C} = 0.0504$$

Hence, I = 0.0504 C + b ------------------- Eq(3)

Substituting any coordinate-point of concentration and inverse path length in Eq(3), the y-intercept is -0.0057.

Hence, the equation of inverse light path length as a function of concentration is formulated as

I = 0.0504 C − 0.0057 ------------------- Eq(4)

The equation could also be formulated in terms of light path length "b" and concentration C as

$$\frac{1}{b} = 0.0504 \text{ C} - 0.0057 \quad \text{------------------- Eq(5)}$$

2-D: The data set of concentration and inverse path length allows the easiest determination of a best-fit line which is essentially Eq(4). This is because the graph of this data set as shown in Figure-5-8 is rectilinear, so an average slope for the whole domain is easily calculated.

The graph of concentration and path length as shown in Figure-5-7 is not rectilinear, so the equation of path length as a function of concentration cannot be formulated as a straight line, rather would be a rational function, as given in Eq(5).

The line of best fit shown in Figure-5-9 is not a good

approximation for the function curve.

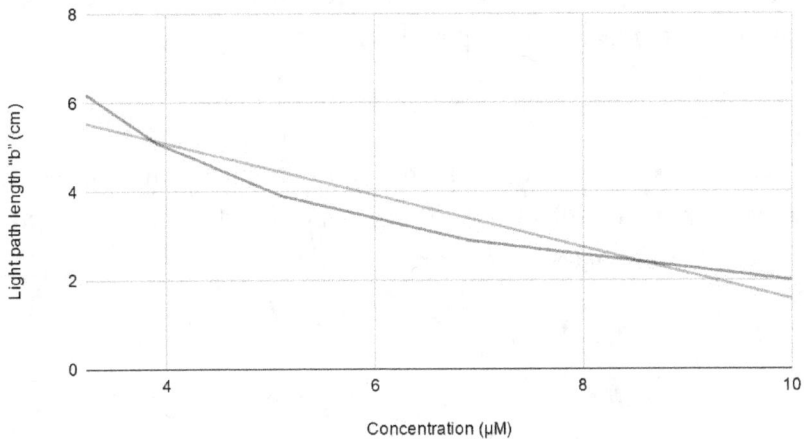

Figure-5-9 Line of best fit for rate of change of light path length with respect to concentration

Procedure to find the concentration of the original solution "A" using Pepsi Blue

The concentration of blue dye#1 in the commercial drink Pepsi Blue is 1 mg/L. The Pepsi Blue can be used to find the true concentration of solution "A".

1-150 mL of Pepsi Blue is put in a 150 mL plastic beaker, which is treated as a stock solution.

2- It is assumed that the original solution of A was prepared in Step-3 of the procedure mentioned earlier.

3-The steps from 5 through 17 of the procedure mentioned earlier are performed to collect data of light path length to be recorded in Table-5-10, and mass of solution in Table-5-11.

Solution	Light path length "b" (cm)
Pepsi Blue as Stock Solution	2.00
A	

Table-5-10 Light path length for solution A and Pepsi Blue

Solution	Mass of test tube and solution (g)	Mass Solution (g)
Pepsi Blue as Stock Solution	1.68	1.33
A		

Table-5-11 Mass of solution A and Pepsi-Blue

4- Mass in grams and volume in mL are nearly the same, so the volume of solution is the same as the quantity of mass. The volume of the solution is recorded in Table-5-12.

Solution	Volume of solution (mL)
Pepsi Blue as Stock Solution	1.33
A	

Table-5-12 Volume of solution A and Pepsi-Blue

5-The molar mass of blue dye is 793g/mol. The concentration of blue dye in Pepsi-Blue is 1 mg/L. So, the number of moles of blue dye in 1 liter of Pepsi-Blue is 1261 µg. Hence, the concentration of blue dye is 1261 µM.

6-The relative concentration of solution A is calculated using the following formula and recorded in Table-5-13.

$$(C_{initial\ stock\ solution}) \times (V_{initial\ stock\ solution}) = (C_{final\ diluted\ solution}) \times (V_{final\ diluted\ solution})$$

Solution	Concentration (μM)	Light path length "b" (cm)	Inverse light path length "1/b" (cm^{-1})
Pepsi Blue as Stock Solution	1261	2.00	0.5
A			

Table-5-13 Concentration, light path length, and inverse of solution A and Pepsi-Blue

7-Graph is drawn for inverse light path length on the y-axis, and concentration on the x-axis to observe the relationship.

To find the concentration of blue-dye in another blue liquid such as Bug juice

Solution A is replaced with Bug juice and steps of "Procedure to find the concentration of original solution A using Pepsi Blue" is followed. Finally, data are measured, calculated, and recorded in Table-5-14.

Solution	Concentration (μM)	Light path length "b" (cm)	Inverse light path length "1/b" (cm^{-1})
Pepsi Blue as Stock Solution	1261	2.00	0.5
Bug-Juice as A			

Table-5-14 Concentration, light path length, and inverse of solution A (BugJuice) and Pepsi-Blue

The concentration is found by using the formula of step 6. Alternatively, the concentration could also be found from the graph plotted in step 7. This graph shows the function curve, where the value of concentration on the x-axis corresponds to the value of light path length on the y-axis. So, the path length of Bug-juice acts as y-coordinate, and the function curve provides the x-coordinate, which is the value of the concentration of Bug-Juice.

Error-Mitigation

1-The comparison of intensity of the color of sample solution with the stock solution is not perfectly matched, rather close enough to be assumed matched. This might cause an error in determining the value of path length, and ultimately the concentration. To get more acceptable data, the average of several trials should be taken.

2-The calculation of the concentration of the sample is not precise enough. Data should be recorded more accurately for better results.

3-The concentration of the stock solution is nearly correct, so the calculation of concentration is only correct relative to the stock solution, but not in absolute terms.

4- Mass in grams and volume in mL are not exactly the same, but nearly the same. It does not change the result much, so can be ignored.

5- The curved bottom of the test tube introduced some error in determining the path length, and volume, but it did not have that big impact, so can be ignored in this study.

Conclusion

Colorimetry is a very quick and efficient way of analyzing colored solutions or any colored substance. Beer-Lambert's law is a linear relationship between the absorption of light and the concentration of the absorbing solution. The absorbance varies directly with the absorptivity coefficient of the solution, light path length, and concentration of the solution, as in Eq(2).

$$A = a\,b\,c$$

In the investigation, the path length and concentration of four sample solutions were determined (Table-5-6) by making their absorbance the same as the stock solution, implying that the absorbance is kept constant. Thus, in Eq(2), absorbance **A** and absorptivity coefficient **a** are constant. This implies that the product of path length **b**, and concentration **c** is a constant, expressed as

b c = Constant

In Table-5-6, it is found that the product of concentration **c**, and light path length **b** of all the four sample solutions and stock solution are constant. Hence, the observation of this investigation is valid, as it follows Beer-Lambert's law.

Using Beer's law the concentration of a solution can be calculated by using the absorbencies. Alternatively, plot a graph of various concentrations, align them to their correct absorbencies, and use a colorimeter to find the concentration of an unknown solution.

The data set of concentration and inverse path length allows the determination of a linear relationship. This is because the graph of this data set as shown in Figure-5-8 is rectilinear, so an equation of the line is easily determined.

The graph of concentration and path length as shown in Figure-5-7 is not rectilinear, so the equation of path length as a function of concentration cannot be formulated as a straight line, rather would be a rational function.

In Figure-5-8, the graph is almost rectilinear, so the function is formulated based on the equation of a line on the Cartesian coordinate of X-Y axis as

y = mx + b, where m is the slope and b is the y-intercept.

In the experiment, the equation of inverse light path length as a function of concentration is formulated as

$$\frac{1}{b} = 0.0504 \, C - 0.0057$$

It was hypothesized at the beginning of this investigation that if the intensity of the color of two solutions is the same, then the product of path length and concentration of the solutions are the same, which turns out to be true, as observed in Table-5-6 and Eq(5).

There is a real-world connection with the concept of colorimetry due to the immense potential of applications ranging from food, textiles, soil, and scientific research. Colorimeters are widely used for monitoring the growth of bacterial or yeast cells in liquid cultures. Colorimeters are used to detect plant nutrients in the soil. They are also widely used to screen chemicals in water such as chlorine, nitrite, fluoride, iron, etc.

Lab 6: Gravimetric Analysis

Goals

To study the gravimetric analysis to determine the concentration of an analyte in a solution.

To find out the amount of phosphate in plant food using gravimetric analysis.

Introduction

Gravimetric analysis is a quantitative method of classical analysis to analyze the composition of a mixture of substances. The element to be determined is isolated in a solid compound of known identity and definite composition. The mass of the element that was present in the original sample can be determined from the mass of this compound.

Gravimetric analysis uses measurements of weight as a technique, wherein the weight of the mixture is compared to the weight of the entire substance to establish percent composition. This is commonly accomplished by dissolving a substance and re-precipitating only one of the components to recover the desired substance.

A typical label on a plant food is labeled with a set of numbers such as X-Y-Z. These numbers imply that the plant food is guaranteed to contain at least X% nitrogen (N), Y% phosphorus (in the form of P_2O_5), and Z% potassium (in the form of K_2O).

The phosphorus can be extracted using magnesium cations in presence of ammonia. The balanced chemical reaction that converts the phosphorus in the plant food to precipitate is

$$PO_4^{3-}(aq) + NH_3(aq) + MgSO_4(aq) + 7H_2O(l) \rightarrow MgNH_4PO_4$$
$$6H_2O(s) + OH^-(aq) + SO_4^{2-}(aq)$$

In this lab, the gravimetric determination of phosphorous is based on the precipitation of magnesium ammonium phosphate hexahydrate from a solution that contains acid phosphate ions, ammonium ions, and magnesium ions. The precipitate forms upon slow neutralization with ammonia of an acidic solution of a phosphate-containing sample.

Hypothesis and variables

Hypothesis: The content of phosphorus (P_2O_5) in the plant food is exactly the same as indicated on the container of plant food. From an empirical point of view, the contents of plant food samples will yield quantitative conversion to "magnesium-ammonium-phosphate-hexahydrate" ($MgNH_4PO_4$ $6H_2O$) and gravimetric determination will accurately and precisely quantify the resulting dry weight of this salt.

Independent Variable: Mass filter paper, 10% $MgSO_4$ solution, 40 mL of ammonia, 1 g of plant food, distilled water

Dependent Variable: Mass precipitate, moles of $MgNH_4PO_4 \cdot 6H_2O$, mass of P_2O_5, % P_2O_5 in the sample

Controlled Variable: Concentration of solution, Filter drying area.

Materials

- Digital balance
- Plastic beaker, 30 mL
- Plastic beaker, 150 mL
- Coffee filters, (4)
- Plastic cups, 12 oz
- Funnel
- Glass beaker, 50 mL
- Glass stir rod
- Magnesium sulfate (**$MgSO_4$**)
- Rubber band
- Distilled water
- Household ammonia (**NH_4OH**)
- Paper towels
- Plant food
- Plates (2)

Procedure

1-A 30 mL plastic beaker was placed on the digital balance and tared.

2-1.0 gram of plant food was added to the beaker. The exact amount of plant food used was recorded in Table-6-1. It was noted that the plant food was dry.

3-The beaker was removed from the balance and 20 mL of distilled water was added to the beaker to dissolve the plant food and was stirred for one minute until the plant food was dissolved.

4-A filter was placed in a funnel and the funnel was placed into the 150 mL plastic beaker. Slowly the solution was poured from step 3 into the filter.

5- The 30 mL beaker was washed with an additional 10 mL of distilled water and poured into the 150 mL beaker from step 4.

6-After the solution had been collected in the 150 mL plastic beaker, the funnel was removed and the filter paper was discarded.

7-A 10% $MgSO_4$ solution (Epsom salt) was prepared by weighing a 2.5 g $MgSO_4$ into a 50 mL beaker. 25 mL of distilled water was added. The $MgSO_4$ was stirred for about two minutes to dissolve it.

8-The $MgSO_4$ solution is poured into the filtrate that was collected in the 150 mL plastic beaker in step 6.

9-40 mL of ammonia was poured into the 50 mL glass beaker. With the glass stir rod, the ammonia solution is stirred and poured into the plant food filtrate.

10-On a dry piece of filter paper, #1 was marked with a pencil. The balance is tared and filter paper is weighed. The mass of filter paper is recorded in Table-6-1.

11-A filter paper was placed over the plastic cup so that it formed a shallow cup. The outside edges were folded down and secured with a rubber band.

12-The stir rod was used to briefly stir the solution from step 9. This solution was carefully poured into the filter. The precipitate was very fine and it took considerable some time to filter.

13-Once the filtering was complete, the filter paper was put on the folded paper towels. The towels had drawn off some liquid and reduced the drying time.

14-When most of the excess moisture was out of the filter paper, it was left on the reaction plate to dry overnight.

15-The lab equipment was cleaned and the experiment was repeated two times more from the beginning in order to average the results.

16-Once the filter paper was dry, the paper and the precipitate were weighed. The value was recorded in Table-6-1.

17-The mass of precipitate was calculated and mentioned in Table-6-1.

18-In addition to procedure data, the following data, and calculations of results were recorded in Table1-1:

- a-% phosphorus (in the form of P_2O_5) listed on the label of the plant food container as shown in Image-6-1.
- b-Moles of $MgNH_4PO_4 \cdot 6H_2O$ from its mass. (Molar mass = 245 g/mol)
- c-Moles of P in the precipitate and its mass.
- d-Mass of P and P_2O_5 ; % P in P_2O_5 .
- e-% P_2O_5 in plant food.
- f-The calculations were done for each trail, and the average was determined.
- g-The percent error was calculated using the average % P_2O_5 in sample and % P_2O_5 labeled on the container of the food plant.

19-All the steps of this procedure are repeated three times for the same sample to record three different trials of observation.

Results

The observed data and calculated results are recorded in Table-6.1.

	Trial 1	Trial 2	Trial 3	Average
Mass plant food (g)	0.99	1.02	1.01	1.01
Mass filter paper (g)	0.88	0.89	0.84	0.87
Mass filter paper + precipitate (g)	1.05	1.08	1.04	1.057
Mass precipitate (g)	0.17	0.19	0.2	0.187
Moles of $MgNH_4PO_4 \cdot 6H_2O$ (mol)	0.000694	0.000775	0.000816	0.000762
Moles of P (mol)	0.000694	0.000775	0.000816	0.000762
Mass of P (g)	0.0215	0.024	0.025	0.0235
Mass of P_2O_5 (g)	0.0493	0.0539	0.058	0.0537
% P_2O_5 in sample	4.98	5.39	5.74	5.37
% P_2O_5 reported on plant food container	6	6	6	6
% Error	17	10.17	4.33	10.5

Table-6-1: Gravimetric experiment data for plant food sample

> Organisch-mineralischer NPK-Dünger 5+6+6
> unter Verwendung von Gesteinsmehl, tierischen Neben-
> produkten (Kat 2 und Kat 3 gemäß VO (EG) Nr. 1069/2009),
> Kaliumdünger aus der Vinasseverarbeitung, Kaliumchlorid,
> Magnesiumoxid, Ammoniumsulfat
>
> 5% N   Gesamtstickstoff
> 6% P_2O_5 Gesamtphosphat
> 6% K_2O Gesamtkaliumoxid
>
> ## Nettomasse: 2,5 kg e
>
> **Hersteller und Inverkehrbringer:**
>
> gpi green partners international
> GmbH & Co. KG
> Rockwoolstr.14 · D-45966 Gladbeck
>
> **Basisland:** Deutschland
>
> **Ausgangsstoffe:**
> Gesteinsmehl, tierische Nebenprodukte (Kat 2 und Kat 3
> gemäß VO (EG) Nr. 1069/2009), Kaliumdünger aus der
> Vinasseverarbeitung, Kaliumchlorid, Magnesiumoxid,
> Ammoniumsulfat
> **Nebenbestandteile:**
> 3,7 % S   Gesamtschwefel
>   2 % S wasserlöslicher Schwefel
> 3,0 % MgO   Gesamtmagnesiumoxid
> 9,9 %   Basisch wirksame Bestandteile
>   (bewertet als CaO)
> 38,0 %   Organische Substanz
>   (bewertet als Glühverlust)
> **Anwendungshilfsmittel:**
> Enthält Mittel zur Staubbindung
> **Stickstoffformen:**
> 2,5 % N   Ammoniumstickstoff

Image-6-1: Contents of plant-food reported on container by seller

Calculations of Results

Calculation for trial-1:

1)- Moles of $MgNH_4PO_4 \cdot 6H_2O$

Since the number of moles of a substance is determined by dividing the mass by its molar mass.

$$Number\ of\ moles = \frac{mass}{molar\ mass}$$

Molar mass of $MgNH_4PO_4 \cdot 6H_2O$ is given as 245 g/mol.

Hence,

moles of $MgNH_4PO_4 \cdot 6H_2O = \frac{mass\ of\ MgNH_4PO_4 6H_2O}{molar\ mass\ of\ MgNH_4PO_4 6H_2O}$

$$= \frac{0.17\ g}{245\ g/mol} = 0.000694\ mol$$

2)- Moles of P

Since 1 mole of $MgNH_4PO_4 \cdot 6H_2O$ contains 1 mole of P, so

moles of P = 0.000694 mol

Alternatively, mole of P can be found as

$$\text{moles of P} = \frac{\text{mass of } MgNH_4PO_4 6H_2O}{\text{molar mass of } MgNH_4PO_4 6H_2O}$$

$$\times \frac{1 \text{ mole P}}{1 \text{ mole } MgNH_4PO_4 6H_2O}$$

$$= \frac{0.17 \text{ g}}{245 \text{ g/mol}} \times \frac{1 \text{ mol}}{1 \text{ mol}} = 0.000694 \text{ mol}$$

3) Mass of P in plant food

The molar mass of P is approximately 31 g /mol.

Mass of a chemical substance is calculated as product of moles and molar mass of that substance.

Mass of P = (moles of P) x (molar mass of P)

= (0.000694 mol) x (31 g /mol) = 0.0215 g

4) Mass of P_2O_5 in plant food

Since there is 2 moles of P in one mole of P_2O_5, so

$$\text{moles of } P_2O_5 = \frac{0.000694 \text{ mol}}{2} = 0.000347 \text{ mol}$$

Alternatively, moles of P_2O_5 can be calculated as

moles of P_2O_5

$$= \frac{mass \text{ of } MgNH_4PO_4 6H_2O}{molar\ mass \text{ of } MgNH_4PO_4 6H_2O}$$

$$\times \frac{1\ mole\ P}{1\ mole\ MgNH_4PO_4 6H_2O} \times \frac{1\ mole\ P_2O_5}{2\ mole\ P}$$

$$= \frac{0.17\ g}{245\ g/mol} \times \frac{1\ mol}{1\ mol} \times \frac{1\ mol}{2\ mol} = 0.000347\ mol$$

Molar mass of P_2O_5 is approximately 142 g/mol.

Mass of P_2O_5 = (moles of P_2O_5) x (molar mass of P_2O_5)

$$= (0.000347\ mol) \times (142\ g/mol) = 0.0493\ g$$

5) % P_2O_5 in plant food

$$\%\ P_2O_5 = \frac{0.0493\ g}{0.99\ g} \times 100 = 4.98\%$$

6) Error% = $\left(\dfrac{Theoretical - Experimental}{Theoretical}\right) \times 100$

$= \dfrac{6 - 4.98}{6} \times 100 = 17\%$

Calculation for trial-2:

1) Moles of $MgNH_4PO_4 \cdot 6H_2O$: $= \dfrac{0.19\ g}{245\ g/mol} \times \dfrac{1\ mol}{1\ mol}$

$= 0.000775\ mol$

2) Moles of P : 0.000775 mol

3) Mass of P : (0.000775 mol) × (31 g /mol) = 0.024 g

4) Mass of P_2O_5 : $\dfrac{0.000775\ mol}{2} \times (142\ g/mol) = 0.055\ g$

5) % P_2O_5 in plant food : $\dfrac{0.055\ g}{1.02\ g} \times 100 = 5.39\%$

6) Error % = $\dfrac{6 - 5.39}{6} \times 100 = 10.17\%$

Calculation for trial-3:

1) Moles of $MgNH_4PO_4 \cdot 6H_2O$: $= \dfrac{0.20\ g}{245\ g/mol} \times \dfrac{1\ mol}{1\ mol} =$ 0.000816 mol

2) Moles of P : 0.000816 mol

3) Mass of P : (0.000816 mol) x (31 g /mol) = 0.025 g

4) Mass of P_2O_5 : $\dfrac{0.000816\ mol}{2}$ x (142 g /mol) = 0.058 g

5) % P_2O_5 in plant food : $\dfrac{0.058\ g}{1.01\ g}$ x 100 % = 5.74%

6) Error % = $\dfrac{6-5.74}{6}$ x 100 % = 4.33%

Discussion/Analysis

Through gravimetric analysis, the percent phosphorus was determined and compared against the amount indicated on the package.

In the gravimetric analysis of the sample, it is investigated that the % P_2O_5 in the sample is 5.37%, while the reported % P_2O_5 on the container of plant food is 6%. There could be a small error in the experiment, but all three trials have shown that % P_2O_5 is less than 6%. It would be significant to analyze the relative quality of the results of the three trials.

For each of the trials, the quality of results is analyzed using different parameters and recorded in Table-6-2.

- For each trail and the average, the % P_2O_5 in the sample is recorded in col-2.
- For each trial, the relative deviation is calculated by subtracting the individual deviation from the average deviation and recorded in col-3.

- For each trial, the individual variance is calculated as squares of the deviations from the average deviation and recorded in col-4.

	% P_2O_5 in sample (col-2)	Deviation from average (col-3)	Individual Variance (col-4)
Trail-1	4.98	0.39	0.1521
Trail-2	5.39	0.02	0.0004
Trail-3	5.74	0.37	0.1369
Average	5.37	0	0

Table-6-2: Quality of results for each trial

It is observed that trial-2 is very close to the average result, but trial-1 and trial-3 are almost equally deviated from the average result to a greater magnitude, as observed in col-3 and col-4 of Table-6-2.

Concepts of precipitation gravimetry were applied in the experiment. Phosphorus from the fertilizer sample was dissolved in water, ammonia, and magnesium sulfate solution. A precipitate, magnesium ammonium phosphate hexahydrate ($MgNH_4PO_4 \cdot 6H_2O$) was formed, which contained all the phosphorus from the fertilizer sample. The precipitate was separated using filtering and weighed. The mass of $MgNH_4PO_4 \cdot 6H_2O$ was divided by its molar mass to determine the number of moles of the precipitate. Using the molar relationship between $MgNH_4PO_4 \cdot 6H_2O$ and phosphorus, the mass of phosphorus was calculated. Similarly, using the molar relationship between phosphorus and P_2O_5, the mass of P_2O_5 was calculated. Finally, the total mass of P_2O_5 was divided by the total sample mass to determine % P_2O_5, which was compared with % P_2O_5 indicated on the container.

The average amount of phosphorus found in plant food sample recorded in Table-6-1 is

5.37 ± 10.5 %.

For the obtained result of 5.37 for % P_2O_5, the percent error of 10.5% is good enough to be accepted. With the observed results, the main objective of quantitatively determining the phosphorous content of plant food through gravimetric analysis was achieved. In all of the trials, the actual amount of phosphorus in plant food is lower than the amount indicated on the container, as observed in Table-6-1.

Error-Mitigation

- It is essential to be careful and not lose the sample solution while transferring from beaker to filter.
- The filtering is a slow process, and it should be ensured that while filtration is underway, any part of the filter should not touch the upper layer of the filtered solution.
- If the precipitate contained in the filter is not completely dry, then the mass of residual water is erroneously considered as the mass of precipitate. Hence, it is to be ensured that the precipitate is fully dried to a constant mass.
- If limiting reagent is not sufficient to complete the reaction, then the mass precipitate turns out to be less than it should be. Hence, an excess of the limiting reagent should be

used. Unused limiting reagent remains unused and does not affect the mass of precipitate.
- Some sources of error could be that filtering was unable to separate all precipitates, and some of the precipitates were lost somehow. It should be ensured that precipitates are properly handled and separated out.
- The filter paper should be of high quality so that all the precipitates are properly separated from the solution.
- The filtration should be done twice so that all the precipitates are collected successfully for the calculation of % P_2O_5.
- The results get better if the too much deviating trial is discarded, and is repeated afresh so that every trial is consistent with valid trials.

Conclusion

Gravimetric analysis is a technique that involves precipitating, selectively, an analyte, and then weighing the precipitate to determine how much analyte is present. The solid precipitate is separated from the surrounding solution by filtering. Phosphorus is one of the most essential components of plant food. To determine how much phosphorus is present in the fertilizer sample, gravimetric analysis through precipitation of magnesium ammonium phosphate hexahydrate was used.

The results of phosphorus were quantitatively determined. The average amount of phosphorus found in the plant food sample recorded in Table-6-1 is 5.37 ± 10.5 %. The experiment showed that the actual amount of phosphorus in plant food is lower than the amount indicated on the container. The results pooled from 3 trials gave a better approximation for the content of phosphorus.

It was hypothesized at the beginning of this investigation that the content of phosphorus (P_2O_5) in the plant food is exactly the same as indicated on the container of plant food, which turns out to be false, as observed in Table-6-1. It is concluded that the actual amount of phosphorus in plant food is lower than the amount indicated on the container, as observed in all of the trails.

The real-world connection of gravimetric analysis is the use of mass measurements to determine the amount of substance that is present. A real-life application would be a gravimetric analysis of water to monitor levels of lead. Other application includes determining the mineral content of drinking water, such as fluoride, calcium, mercury, calcium, etc. Gravimetric analysis is also used to calculate the gold content in jewelry, fat content in milk, etc. Another important usage is the isolation of sodium hydrogen bicarbonate from a mixture of carbonate and bicarbonate.

7 Lab 7: Bond and Molecular Polarity

Goals
To predict bond polarity using electronegativity values
To indicate polarity with a polar arrow or partial charges
To rank bonds in order of polarity
To predict molecular polarity using bond polarity and molecular shape

Introduction

Electronegativity is the power of an atom in a molecule to attract electrons to itself. It is quantified in terms of Pauling scale value. If an atom has a high electronegativity, then it also has a high electron affinity. Electronegativity increases from the left group to the right group, and from bottom to top in a group of the periodic table.

Polarity is a separation of electric charge leading to a molecule or its chemical groups having an electric dipole moment, with a negatively charged end and a positively charged end. Bond polarity exists when two bonded atoms unequally share electrons, resulting in a negative and positive end.

Polar molecules must contain polar bonds due to a difference in electronegativity between the bonded atoms. A polar molecule with two or more polar bonds must have a geometry, which is asymmetric in at least one direction so that the bond dipoles do not cancel each other. Polar molecules interact through dipole-dipole intermolecular forces and hydrogen bonds. Polarity underlies a number of physical properties including surface tension, solubility, and melting and boiling points.

Using net electronegativity difference it is predicted whether a given bond is non-polar, polar covalent, or ionic. The greater the difference in electronegativity the more polar the bond. If the difference in electronegativity between two atoms in a bond is above 0.5, then the bond is a polar covalent bond. If the difference in electronegativity is less than 0.5, then the bond is a nonpolar covalent bond. Similarly, if the difference in electronegativity is above 1.7, then the bond is an ionic bond.

Hypothesis and variables

Hypothesis: Molecular polarity is the vector sum of bond polarity vectors of each atom in the molecule.

Independent Variable: Electronegativity difference of bonds, shape of the molecule, central and outside atoms in the molecule.

Dependent Variable: Bond polarity, dipole moment, molecular polarity.

Controlled Variable: None.

Materials:

- Access to website:
  http://phet.colorado.edu/en/simulation/molecule-polarity

Procedure:

1. Website URL

   < http://phet.colorado.edu/en/simulation/molecule-polarity > was accessed, and the applet was executed. The simulation of electronegativities and polarity were explored.

2. Two Atoms System was explored to find out how changing the electronegativity difference affects the

   a. bond dipole

   b. partial charges

   c. bond character

   d. electrostatic potential

   e. electron density

   f. response to an electric field

3. The observations and results were recorded in Table-7-1. The descriptions were mentioned near the data table.

4. Three Atoms System was explored to find out how changing the electronegativity difference affects the

   a. bond dipole

   b. molecular dipole

   c. partial charges

   d. electric field

5. In the Real molecules simulation, all the molecules were explored with respect to bond dipoles, molecular dipole, partial charges, electrostatic potential, and electron density.

   For each of the molecules, the atoms with partial negative charges, atoms with negative electrostatic potential, and atoms with more electron density were recorded in the Table-7-3.

a. In addition, the electronegativity differences were calculated for all bonds of the atoms in the molecule.
   b. The shapes of molecules were analyzed, and reasons for molecular dipole were determined.
   c. The molecules were classified according to their bond types. Molecules with nonpolar covalent bonds were recorded in Table-7-4, and molecules with polar covalent bonds were recorded in Table-7-5.
6. The overall process of determining a molecule possessing a specific bond type was analyzed and a flowchart was developed in Figure-7-6.

Results and Descriptions

Two atoms system

A bond dipole is the separation of charges between bonded atoms due to differences in electron density caused by electronegativity difference. The bond dipole is shown by an arrow starting from the less electronegative atom and pointing towards the more electronegative atom with a cross at the tail.

The partial charges are the strength of equal and opposite charges on two atoms. In a polar covalent bond, an atom having higher electronegativity in a bond gets partial negative denoted by δ^- charges, while a less electronegative atom gets partial positive charges denoted by δ^+.

The bond types between atoms are ionic, polar covalent, or nonpolar covalent. If the difference in electronegativity between two atoms in a bond is above 0.5, then the bond is a polar covalent bond. If the difference in electronegativity is less than 0.5, then the bond is a nonpolar covalent bond. Similarly, if the difference in electronegativity is above 1.7, then the bond is an ionic bond.

Electrostatic potential surrounding the atom is the restored energy which equals the amount of work needed to move a unit of charge from a reference point to a specific point inside the field.

Electron density is the measure of the probability of an electron being present at a specific location.

A field effect is the polarization of a molecule through space. The effect is a result of an electric field produced by charge localization in a molecule.

	Effect due to change in electronegativity difference
Bond dipole	1-The atom with larger electronegativity has more pull for the bonded electrons than the pull of atom with smaller electronegativity; 2-The net dipole is in a direction towards atom having stronger pull; 3-The greater the difference in the two electronegativities, the larger is the dipole. 4-If both atoms are equally electronegative, the dipole ceases to exist.
Partial charges	1-If the electronegativities of two atoms are equal, then there are no partial charges. 2-The greater the difference in the two electronegativities, the larger is the dipole. 3-More electronegative atom acquires negative charge, whereas other atom acquires positive charge. 4- In a polar covalent bond, an atom having higher electronegativity in a bond gets partial negative denoted by δ^- charges, while less electronegative atom gets partial positive charges denoted by δ^+.
Bond character	1-As the electronegativity difference decrease, the bond becomes covalent. 2- As the electronegativity difference increases, the bond becomes polarized. 3-The bond character is strongest covalent, when the electronegativity difference is zero or very low.

	4- The bond character is strongest ionic, when the electronegativity difference is greatest. 5- Electronegativity differences can be used to predict if a bond is covalent, polar covalent or ionic
Electrostatic potential	1-Electrostatic potential is largest when electronegativity difference is greatest. 2-If the electronegativities of two atoms are equal, then there is no electrostatic potential. 3- The greater the difference in the two electronegativities, the larger is the electrostatic potential. 4- More electronegative atom acquires negative electrostatic potential, whereas another atom acquires positive electrostatic potential.
Electron density	1-The greater the electronegativity difference between the two atoms, the more the electron density is pulled toward the more electronegative atom. 2- In a bond, the more electronegative element has a greater share of the electrons, and a partial negative charge to reflect this greater electron density. 3- The less electronegative element has a partial positive charge to reflect the lack of electron density. 4- If two atoms have exactly the same electronegativity, then the electron density of the two shared electrons in the bond are shared equally.
Response to an electric field	1-Polar molecules orient themselves in the presence of an electric field with the positive ends of the molecules being attracted to the negative plate while the negative ends of the molecules are attracted to the positive

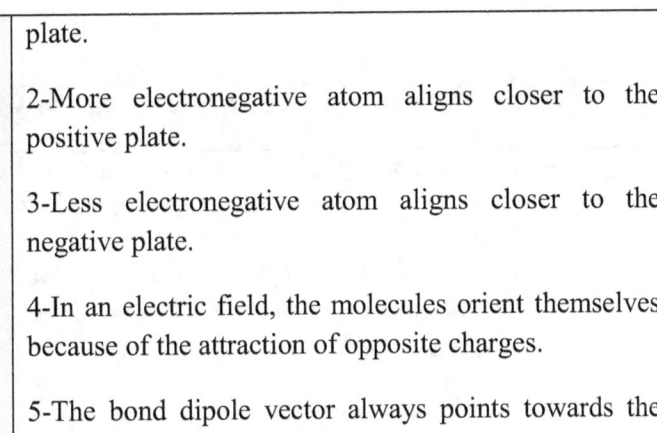	plate. 2-More electronegative atom aligns closer to the positive plate. 3-Less electronegative atom aligns closer to the negative plate. 4-In an electric field, the molecules orient themselves because of the attraction of opposite charges. 5-The bond dipole vector always points towards the positive plate.

Table-7-1: Two atoms system - effect due to change of electronegativity difference

Three atoms system

In the three atoms system, there are two bond dipoles between central atom B and each of two outside atoms A and C. Each of the bond dipoles is the separation of charges between the bonded atoms due to differences in electron density caused by electronegativity difference. The two bond dipoles are shown by two different arrows starting from the less electronegative atom and pointing towards the more electronegative atom with a cross at the tail.

In three atom systems, the partial charges are the strength of charges for the whole molecule. In a polar molecule, partial charges are shown for each of three atoms, where an atom having higher electronegativity gets partial negative denoted by δ^- charges, while a less electronegative atom gets partial positive charges denoted by δ^+.

	Effect due to change in electronegativity difference
Bond dipole	1-There are two bond dipoles between central atom B and each of the two outside atoms A and C. 2-The two bond dipoles are independent of each other and just depend upon electronegativity difference between bonded atoms. 3-The atom with larger electronegativity has more pull for the bonded electrons than the pull of atom with smaller electronegativity; 4-The greater the difference in the two electronegativities, the larger is the dipole. 5-If both atoms are equally electronegative, the dipole ceases to exist.
Molecular dipole	1-Molecular dipole is the vector sum of two bond polarity vectors. 2-The molecular dipole vector shows the magnitude and direction of the net polarity of the whole molecule. 3- The length of the molecular dipole vector indicates the strength of polarity and direction of vector points from positive to the negative charge of the molecule. 4-There is the symmetrical shape of three atoms aligned on vertices of an equilateral triangle, and atoms possess the same electronegativity the symmetrical shape cancels the dipole and the molecule is nonpolar. 5-Even if the shape of the molecule is symmetrical, but outside atoms A and C do not have the same electronegativity, then the molecular dipole exists, and

	the molecule is polar.
Partial charges	1-The partial charges show the strength of charges on an atom. 2-If the electronegativities of three atoms are equal, then there are no partial charges. 3-The greater the molecular dipole, the larger is the partial charges. 4-The molecular dipole vector points towards the center of mass of partial negative charged atoms.
Response to an electric field	1-The molecule orients itself, such that the molecular dipole vector points towards a positive electric field. 2-Polar molecules orient themselves in the presence of an electric field with the positive ends of the molecules being attracted to the negative plate while the negative ends of the molecules are attracted to the positive plate. 3-The partial negative charged atom aligns closer to the positive plate of field, and the partial positive charged atom aligns closer to the negative plate. 4-In an electric field, the molecule orients itself because of the attraction of opposite charges.

Table-7-2: Two atoms system - effect due to change of electronegativity difference

In the three atoms system, a field-effect is the polarization of a molecule through space, where the molecule orients itself such that molecular dipole points directly towards a positive plate.

Real molecules system

For each of the molecules, the atoms with partial negative charges, atoms with negative electrostatic potential, and atoms with more electron density are recorded in Table-7-3.

	Atom with partial negative charge (δ^-)	Negative electrostatic potential	Atom with more electron Density
H_2	None	Uniform	Uniform
N_2	None	Uniform	Uniform
O_2	None	Uniform	Uniform
F_2	None	Uniform	Uniform
HF	F	F	F
H_2O	O	O	O
CO_2	O1, O2	O1, O2	O1, O2
HCN	N	N	N
O_3	O1, O3	O1, O3	O1, O3
NH_3	N	N	N
BF_3	F1, F2, F3	F1, F2, F3	F1, F2, F3
CH_2O	O	O	O
CH_4	C	C	C
CH_3F	F	F	F

CH_2F_2	F1, F2	F1, F2	F1, F2
CHF_3	F1, F2, F3	F1, F2, F3	F1, F2, F3
CF_4	F1, F2, F3, F4	F1, F2, F3, F4	F1, F2, F3, F4
$CHCl_3$	Cl1, Cl2, Cl3	Cl1, Cl2, Cl3	Cl1, Cl2, Cl3

Table-7-3: Real molecules system – partial charges, electrostatic potential, and electron density

The electronegativity differences are calculated for all bonds in the molecule. The shapes of molecules are analyzed, and reasons for molecular dipole are determined. The molecules are classified according to their bond types. Molecules with nonpolar covalent bonds were recorded in Table-7-4, and molecules with polar covalent bonds were recorded in Table-7-5.

There is no ionic molecule observed in the investigation.

Nonpolar covalent molecules due to molecular dipole being zero or too low are

$$F_2, CO_2, CH_4, H_2, N_2, O_2, F_2, CF_4, BF_3, BH_3, HCN$$

	ΔEN Atoms	ΔEN Atoms	ΔEN Atoms	ΔEN Atoms	Shape / Angle of polar bonds	Reason of no/low dipole
F_2	F-F=0	F=4			Linear	ΔEN=0
CO_2	C-O=1	C-O=1			Linear triatomic (180°)	Polarity cancels out due to symmetry
CH_4	C-H=0.4	C-H=0.4	C-H=0.4		Tetrahedral (109°)	Polarity cancels out due to symmetry
H_2	H-H=0				Linear	ΔEN=0
N_2	N-N=0				Linear	ΔEN=0
O_2	O-O=0				Linear	ΔEN=0
F_2	F-F=0				Linear	ΔEN=0
CF_4	C-F=1.5	C-F=1.5	C-F=1.5	C-F=1.5	Tetrahedral (109°)	Polarity cancels out due to symmetry
BF_3	B-F=2	B-F=2	B-F=2		Trigonal Planar (120°)	Polarity cancels out due to symmetry
BH_3	B-H=0.1	B-H=0.1	B-H=0.1		Trigonal Planar	Polarity cancels

					(120°)	out due to symmetry
HCN	H-C=0.4	C-N=0.5			Linear triatomic (180°)	Polarity is too low due to symmetry towards N atom

Table-7-4: Real molecules system – Nonpolar covalent bond

Polar covalent bond molecules due to molecular dipole being greater than 0.5 are

H_2O, CH_3F, CH_2F_2, CHF_3, $CHCl_3$, CH_2O, NH_3, O_3, HF

	ΔEN Atoms	ΔEN Atoms	ΔEN Atoms	ΔEN Atoms	Shape / Angle of polar bonds	Reason of high dipole
H_2O	O-H=1.4	O-H=1.4			Tetrahedral	Symmetrical shape but uneven charge distribution; oxygen is more electronegative than hydrogen; Net dipole points towards O atom.
CH_3F	C-H=0.4	C-H=0.4	C-H=0.4	C-F=1.5	Tetrahedral	Symmetrical shape but

						uneven charge distribution; Outside atom F is more electronegative, so polarity does not cancel out; Net dipole points towards F atom.
CH_2F_2	C-H=0.4	C-H=0.4	C-F=1.5	C-F=1.5	Tetrahedral	Symmetrical shape but uneven charge distribution; Outside atom F is more electronegative, so polarity does not cancel out; Net dipole points towards the middle of two F atoms.
CHF_3	C-H=0.4	C-F=1.5	C-F=1.5	C-F=1.5	Tetrahedral	Symmetrical shape but uneven charge distribution; Outside atom F is more electronegative, so polarity does not cancel out; Net dipole points towards centroid of three F atoms.
$CHCl_3$	C-	C-	C-	C-	Tetrahedral	Symmetrical shape but

							uneven charge distribution; Outside atom F is more electronegative, so polarity does not cancel out; Net dipole points towards centroid of three Cl atoms.
CH_2O	C-H=0.4	C-H=0.4	C-O=1			Trigonal planar	Symmetrical shape but uneven charge distribution; oxygen atom is more electronegative than either carbon or hydrogen atom. Net dipole exists pointing towards O atom.
NH_3	N-H=0.9	N-H=0.9	N-H=0.9			Trigonal pyramidal	The polarity of N-H does not cancel out; Net dipole exists and points towards N atom.
O_3	O-O=0	O-O=0	O-O=0			Bent	Electron density shows resonance pattern; Dipole exists;

HF	H-F=1.9				Linear diatomic	ΔEN =1.9; H is nonmetallic, so not ionic bond; Molecular dipole exists and points towards N;

Table-7-5: Real molecules system – Polar covalent bond

Discussion/Analysis

It was observed that there are several polar covalent compounds, nonpolar compounds, but no ionic compounds. If the difference in electronegativity between two atoms in a bond is above 0.5, then the bond is a polar covalent bond. If the difference in electronegativity is less than 0.5, then the bond is a nonpolar covalent bond. Similarly, if the difference in electronegativity is above 1.7, then the bond is an ionic bond.

In the two atoms system, the polar bond between two atoms implies that their electronegativity difference is greater than zero, wherein a single bond dipole between two atoms determine the molecular polarity. In the three atoms system, the vector sum of bond dipoles determines the molecular polarity.

The two atoms system is linear, so the molecular dipole is an algebraic difference of electronegativities of two atoms. But in three atom systems, the shape of a molecule plays an important role, where the bond dipole between the central atom and each of the outside atoms are vectorially added to find the resultant vector for the molecular dipole. In the three atoms system, the electronegativities of outside atoms also play an important role in determining the molecular dipole.

Polar molecules occur when two atoms do not share electrons equally in a covalent bond. A dipole forms, with part of the molecule carrying a slight positive charge and the other part carrying a slight negative charge. This happens when there is a difference between the electronegativity of each atom. An extreme difference forms an ionic bond, while a lesser difference forms a polar covalent bond.

When molecules share electrons equally in a covalent bond there is no net electrical charge across the molecule. In a nonpolar covalent bond, the electrons are evenly distributed. Nonpolar molecules are formed when atoms have the same or similar electronegativity. In general, if the electronegativity difference between two atoms is less than 0.5, the bond is considered nonpolar, even though the only truly nonpolar molecules are those formed with identical atoms.

The O_3 molecule consists of three oxygen atoms, one single coordinate covalent bond, and one double covalent bond. The two O-O that share the double covalent bond are nonpolar as there is no electronegativity between these atoms of the same element, sharing the same number of electrons. However, the distribution of the electrons along the 3 oxygen atoms is uneven, as the central atom must share electrons with the two atoms on either side of it, whereas the outer atoms must share only with the central one. Due to the electrons being taken away from the central atom, this atom is more deprived of electrons than the other two-oxygen atoms. As a result, the central atom has a formal charge of +1 while the other atoms each have a formal charge of -½. Since the molecule is bent in terms of geometry, the result of this sharing of electrons results in a dipole force across the entirety of the ozone molecule.

The electron density is uniformly distributed across both the oxygen atoms of the oxygen molecule. The electron density in the ozone molecule is non-uniformly distributed. The two outside oxygen atoms have higher electron densities than the central oxygen atom. The two outside oxygen atoms have -1/2 formal charge, which is due to the fact that electron density moves around the molecule continually and thus both atoms exhibit the same properties. Because of this, the Ozone exhibit a resonance behavior in its structure.

The electrostatic potential of oxygen is almost neutral and uniformly distributed across the whole molecule. In ozone, the electrostatic potential is positive for the central oxygen atom, but for two outside oxygen atoms, the electrostatic potential is slightly negative. The two outside atoms possess similar electrostatic potential, but different than the central oxygen atom due to the resonance pattern.

The molecules of methane, fluoromethane, and chloroform have a tetrahedral symmetrical shape. The methane molecule is nonpolar because of its symmetrical shape possessing the same outside atoms, but fluoromethane, and chloroform are polar because outside atoms are different and bond dipoles do not get canceled out. This effect is also observed among the three compounds with respect to electrostatic potential and electron density.

In methane, the electron density around two diagonally opposite H atoms is slightly more than the other two H atoms. In fluoromethane, the electron density is substantially high around the F atom than three of H atoms. In chloroform, the electron density is substantially low around the H atom as compared to C and Cl atoms.

In methane molecule, the electrostatic potential around two diagonally opposite H atoms is more negative than the other two H atoms. In fluoromethane, the electrostatic potential around the F atom is highly negative than C and H atoms. In chloroform, the electrostatic potential around the H atom is highly positive as compared to C and Cl atoms.

The process of determining molecular dipole involves the following steps:

1-If the difference in electronegativity between two atoms in a bond is above 0.5, then the bond is a polar covalent bond. If the difference in electronegativity is less than 0.5, then the bond is a nonpolar covalent bond. Similarly, if the difference in electronegativity is above 1.7, then the bond is an ionic bond.

2-In a polar covalent bond, an atom having higher electronegativity in a bond gets partial negative denoted by δ^- charges, while a less electronegative atom gets partial positive charges denoted by δ^+.

3-In an ionic bond, an atom with higher electronegativity gaining an electron does not get a partial negative charge, rather gets a full formal negative charge. Similarly, an atom losing an electron in an ionic bond gets a full formal positive charge.

4-a-The movement of electrons from a partial positive charged atom to a partial negative charged atom is shown by an arrow pointing towards a partial negative charged atom with a cross at the tail. The length of the arrow is proportional to the magnitude of bond polarity, which is equal to the electronegativity difference.

4-b- The arrow showing the movement of electrons is considered as a vector showing the bond polarity. Each bond polarity of a molecule is shown by a single arrow treated as a vector. If all the bond polarities of a molecule are added together, then the resultant vector shows the net polarity of the molecule.

4-c- The resultant vector indicates the magnitude and direction of a molecule's polarity, pointing towards the negative charge of the molecule.

5- In a polar covalent bond, an atom with partial negative charges has higher electron density, while an atom with partial positive charges has lesser electron density. This implies that if a negative charge is unevenly distributed around the atoms of a molecule, then the bond is polar.

6- If the outside atoms are evenly distributed around the central atom, then the symmetrical shape might cancel the dipole and the molecule is nonpolar.

Symmetrical shapes are:

- Trigonal planar
- Tetrahedral
- Linear triatomic

Non-symmetrical shapes are:

- Bent
- Pyramidal
- Diatomic

7- If the outside atoms are not evenly distributed causing the shape to be non-symmetrical, then the dipole might exist, and the molecule is polar.

Flowchart

The overall process of determining a molecule possessing a specific bond type is analyzed and a flowchart is developed in Figure-7-6.

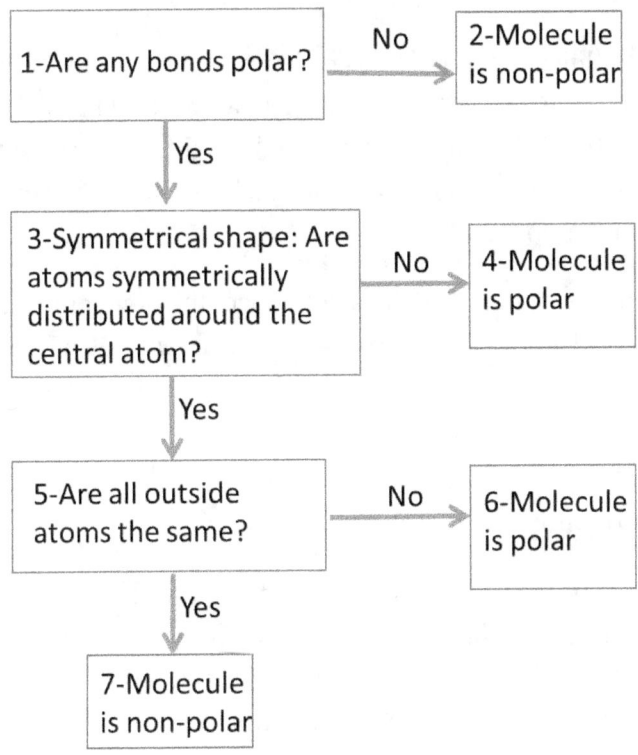

Figure-7-6: Flowchart for determining the bond types

Use-cases

CO_2 is completely non-polar due to linear symmetrical structure. Bond polarities cancel each other, so there is no dipole.

BF₃ with boron in the center has a trigonal planar symmetrical shape, where the fluorines are equidistant from each other at 120°. The electronegativities show a large difference in the B-F bond resulting in a large bond polarity in each bond. However, if the bond polarity vectors are added up, the resulting vector sums up to zero. Hence, the bond polarities cancel due to symmetry, so there is no dipole.

HCN is a linear triatomic. The electronegativity differences in bonds, and thus the bond polarities are different. Adding the individual bond polarity vector together gives a resultant vector pointing towards a more electronegative atom, oxygen. Hence, HCN is a polar molecule.

Thy symmetry in this case does not cancel the bond polarities since different outside atoms result in different electronegativity differences.

Error mitigation

Two atoms system does not have a dependency, as only one bond dipole is involved, but in three atoms system, there might be a dependency of the second bond dipole. For an error-free investigation, many possible combinations of electronegativities of three atoms should be observed.

In real molecule investigations, there are some exceptions due to specific behavior and the result is not as expected. For instance, an ozone molecule exhibits resonance behavior and is a polar molecule, although the symmetrical shape and same electronegativity of oxygen atom should cancel out the bond dipole. For the error-free investigation, all the molecules should be known for their particular or exceptional behaviors.

The simulation does not show the detailed process and concepts, which might lead to an error. For instance, the process could be pictorially shown, how the bond dipole contributes and leads to a molecular dipole. The concepts should be well known for error-free results.

Conclusion

In the two atoms system, the polar bond between two atoms implies that their electronegativity difference is greater than zero, wherein a single bond dipole between two atoms determines the molecular polarity. In the three atoms system, the vector sum of bond dipoles determines the molecular polarity.

The two atoms system is linear, so the molecular dipole is an algebraic difference of electronegativities of two atoms. But in three atom systems, the shape of a molecule plays an important role, where the bond dipole between the central atom and each of the outside atoms are vectorially added to find the resultant vector for the molecular dipole. In the three atoms system, the electronegativities of outside atoms also play an important role in determining the molecular dipole.

Bond polarity affects the polarity of a molecule. If the negative charge is evenly distributed around the molecule, then the molecule is not polar. However, if the negative charge is not evenly distributed, then the molecule is polar.

Electronegativity difference gives the magnitude of the bond polarity shown by the arrow starting from the lower electronegativity atom and pointing to the higher electronegative atom. The length of the arrow is proportional to the magnitude of bond polarity, which is equal to the electronegativity difference.

The arrow showing the movement of electrons is considered as a vector showing the bond polarity. Each bond polarity of a molecule is shown by a single arrow treated as a vector. If all the bond polarities of a molecule are added together, then the resultant vector shows the net polarity called dipole of a molecule. The resultant vector indicates the magnitude and direction of a molecule's polarity, pointing towards the negative charge of the molecule.

If the outside atoms are evenly distributed around the central atom, then the symmetrical shape might cancel the dipole and the molecule is nonpolar.

If the outside atoms are not evenly distributed causing the shape to be non-symmetrical, then the dipole might exist, and the molecule is polar.

It is also observed that in polar covalent compounds, atom possessing partial negative charge also has negative electrostatic potential and higher electron density.

It was hypothesized at the beginning of this investigation that in polar covalent compounds, an atom possessing partial negative charge also has negative electrostatic potential and higher electron density, which turns out to be true, as observed in Table-7-3.

The real-world connection of bond and molecular polarity is the understanding of substances around us. The concept of polarity helps understand the states of matter, shapes of molecule, specific behavior of compounds, solubility, conductivity, etc.

8 Lab 8: Intermolecular Forces

Goal

To evaluate factors affecting the boiling and melting points of selected compounds.

Introduction

There are two types of forces active in a molecule, intramolecular and intermolecular. Intramolecular forces are the forces that hold atoms together within a molecule. Intermolecular forces are forces that exist between molecules.

Types of intramolecular forces of attraction:

1-Ionic bond: This bond is formed by the complete transfer of valence electron(s) between atoms. It is a type of chemical bond that generates two oppositely charged ions. In ionic bonds, the metal loses electrons to become a positively charged cation, whereas the nonmetal accepts those electrons to become a negatively charged anion.

2-Covalent bond: This bond is formed between atoms that have similar electronegativities—the affinity or desire for electrons. Because both atoms have a similar affinity for electrons and neither has a tendency to donate them, they share electrons in order to achieve octet configuration and become more stable.

3-A nonpolar covalent bond: It is formed between the same atoms or atoms with very similar electronegativities—the difference in electronegativity between bonded atoms is less than 0.5.

4-Metallic bond: This type of covalent bonding specifically occurs between atoms of metals, in which the valence electrons are free to move through the lattice. This bond is formed via the attraction of the mobile electrons—referred to as a sea of electrons—and the fixed positively charged metal ions. Metallic bonds are present in samples of pure elemental metals, such as gold or aluminum, or alloys, like brass or bronze.

The relative strength of intramolecular forces increases in the order of nonpolar covalent bond, polar covalent bond, ionic bond, and metallic bond.

Intermolecular forces are much weaker than the intramolecular forces of attraction but are important because they determine the physical properties of molecules like their boiling point, melting point, density, and enthalpies of fusion and vaporization.

Types of intermolecular forces that exist between molecules:

1-Dipole-dipole interactions: These forces occur when the partially positively charged part of a molecule interacts with the partially negatively charged part of the neighboring molecule.

2-Hydrogen bonding: This is a special kind of dipole-dipole interaction that occurs specifically between a hydrogen atom bonded to either oxygen, nitrogen, or fluorine atom. The partially positive end of hydrogen is attracted to the partially negative end of the oxygen, nitrogen, or fluorine of another molecule. Hydrogen bonding is a relatively strong force of attraction between molecules, and considerable energy is required to break hydrogen bonds.

3-London dispersion forces, under the category of van der Waal forces: These are the weakest of the intermolecular forces and exist between all types of molecules, whether ionic or covalent—polar or nonpolar. The more electrons a molecule has, the stronger the London dispersion forces are.

The relative strength of intermolecular forces increases in the order of London dispersion forces, dipole-dipole attraction, and hydrogen bonding.

Hypothesis and variables

Hypothesis: The melting or boiling points increases with increasing molar mass.

Independent Variable: Data-set A, Data-set B, Data-set C.

Dependent Variable: Melting/boiling point.

Controlled Variable: None.

Materials

Data Sets (Data-set A in Table-8-1, Data-set B in Table-8-2, Data-set C in Table-8-3)

Procedure

1. A data table was constructed for each set of given data. Data-set A was recorded in Table-8-1, data-set B in Table-8-2, and data-set C in Table-8-3.

2. The data was graphed. Figure-8-4 was for data-set A, Figure-8-5 for data-set B, and Figure-8-6 for data-set C. The trend lines were included where appropriate and discrepant points were indicated.

3. The observed trends were explained in melting or boiling points based on bonding and intermolecular forces.

4. The discrepant points were explained.

Data Sets:

Given data set-A is recorded in Table-8-1.

Compound	Melting point
HI	-50.8 °C
HBr	-88.5 °C
HCl	-114.8 °C
HF	-81.3 °C

Table-8-1: Data-set A

Given data set-B is recorded in Table-8-2.

Compound	Boiling point
H_2Te	-2 °C
H_2Se	-41.5 °C
H_2S	-60.7 °C
H_2O	100 °C

Table-8-2: Data-set B

Given data set-C is recorded in Table-8-3.

Compound	Boiling point
C_2H_6	-88 °C
C_4H_{10}	-42 °C
C_6H_{14}	70 °C
C_8H_{18}	126 °C
$C_{10}H_{22}$	174 °C

Table-8-3: Data-set C

Results and Analysis:

The given data-set A is graphed in Figure-8-4.

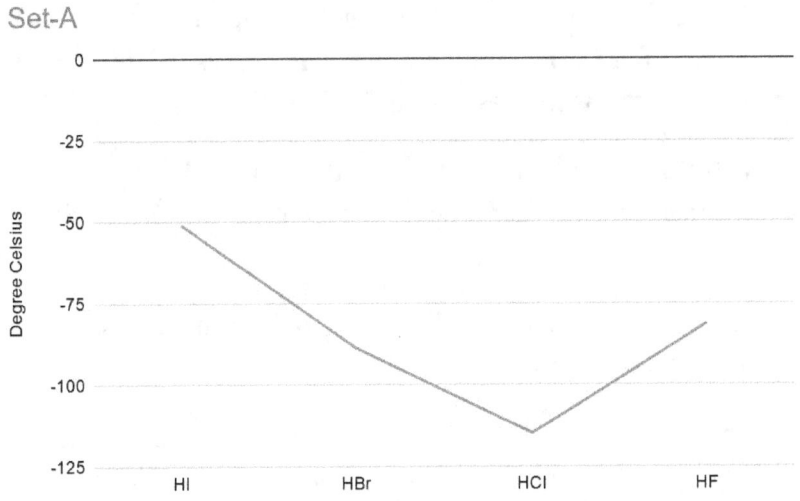

Figure-8-4: Graph of data-set A

Analysis of data-set A:

(a) The atomic mass of the halogen decreases in the order I, Br, Cl. A lighter element is melted more easily than a heavy one because the lighter molecule moves around a lot faster than the heavier molecule and thus it requires less energy to separate molecules into the liquid phase.

Hence, the melting points are in the order of HCl < HBr < HI.

(b) All the given molecules viz. HF, HCl, HBr, and HI have permanent dipoles. Hence, all of them possess dipole-dipole and London forces. HF in addition to dipole-dipole and London forces also has hydrogen bonding.

(c) Electronegativity of Cl, Br and I is in the order: Cl > Br > I. Therefore, polar character and hence dipole-dipole interactions should be in the order HCl > HBr > HI. But melting points are found to be in the order HCl < HBr < HI. This shows that London forces are predominant. This is because London forces increase as the number of electrons in the molecule increases. In this case, the number of electrons increases from HCl to HI.

(d) Due to the very high electronegativity of F, HF is most polar, and also there is hydrogen bonding present in it. Hence, it has the second-highest melting point.

(e) In the given compounds, HI has the highest molar mass. Hence it has the highest melting point.

As observed in the graph of Figure-8-4, the melting point increases with molar mass except for HF, which has a lower molar mass than HCl but a higher melting point. It is due to hydrogen bonding.

The given data-set B is graphed in Figure-8-5.

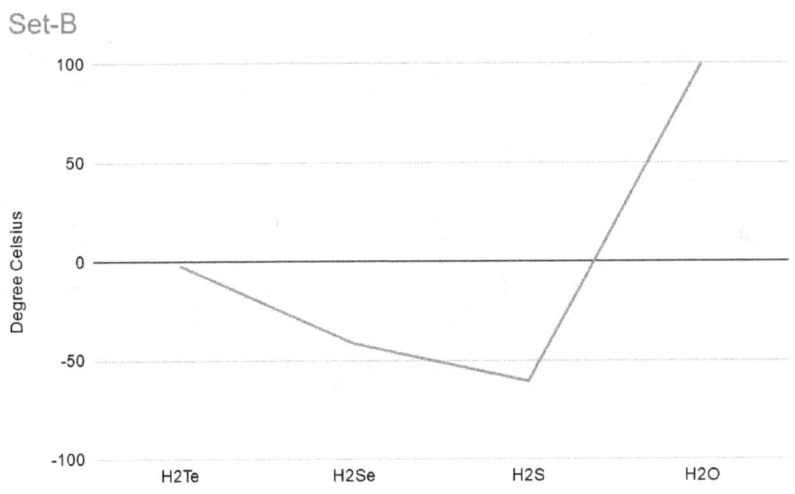

Figure-8-5: Graph of data-set B

Analysis of data-set B:

(a) Moving down group VI the increasing atomic mass requires more energy to overcome its molecular sluggishness and vaporize.

(b) The general increase in boiling point from H_2S to H_2Te is caused by increasing London forces between molecules due to an increasing number of electrons.

(c) The anomalously high boiling point of H_2O is caused by the hydrogen bonding between these molecules in addition to their London forces. The additional forces require more energy to break and so have higher boiling points.

As observed in the graph of Figure-8-5, the boiling point increases with molar mass except for H_2O, which has a lower molar mass than H_2S but a higher boiling point. It is due to hydrogen bonding.

The given data-set C is graphed in Figure-8-6.

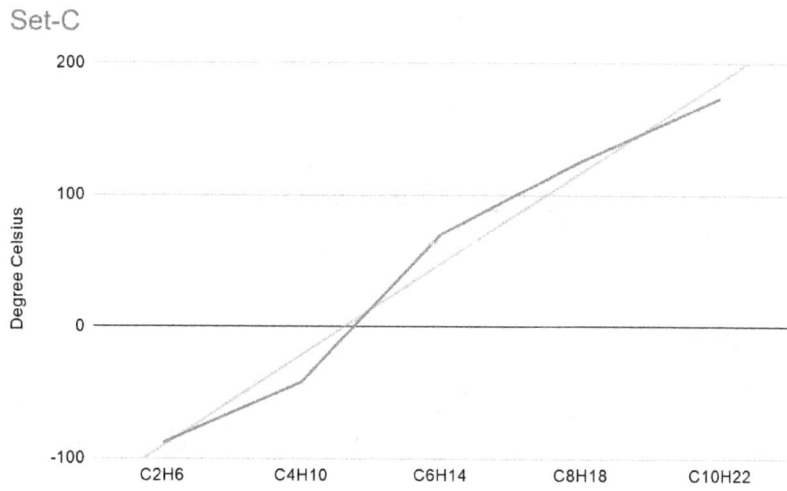

Figure-8-6: Graph of data-set C

Analysis of data-set C:

(a) The longer the carbon chain and the bigger the molar mass, the stronger the London dispersion forces will be.

(b) The shape of the molecule occupying the larger the surface area, the stronger the London dispersion forces.

(c) An interesting thing takes place with butane C_4H_{10}. Butane has two constitutional isomers called butane and isobutane. Butane is a straight-chain hydrocarbon, while isobutane is a branched-chain hydrocarbon. Because the branched-chain isomer has a smaller surface area, you could predict a lower boiling point when compared with straight-chain butane.

(d) For given hydrocarbons, the boiling points increase with increasing molar mass.

(e) All of these compounds are nonpolar and only have London dispersion forces: the larger the molecule, the larger the dispersion forces, and the higher the boiling point. The ordering from lowest to highest boiling point is, therefore, $C_2H_6 < C_4H_{10} < ... < C_{10}H_{22}$

As observed in the graph of Figure-8-6, the boiling point increases with molar mass for all hydrocarbons. All of these compounds are nonpolar and only have London dispersion forces, so the trend line fits closely to the curve of increasing boiling points with molar mass.

Error mitigation

To investigate intermolecular forces better, some more elements from the same group should be included in the investigation to check if some more discrepant points exist.

The reasons for discrepant points should be known and the rest of the data should closely align the trend line. This would improve our observations and understanding.

The data set should be more comprehensive to understand intermolecular forces better.

Conclusion

In this lab, intermolecular forces are investigated while studying the melting/boiling points of the same group elements. There are three different types of intermolecular forces - dipole-dipole attraction, hydrogen bonding, and London dispersion forces. The relative strength of intermolecular forces increases in the order of London dispersion forces, dipole-dipole attraction, and hydrogen bonding.

In Figure-8-4, it is observed that melting points for halogens increase with increasing molar mass of halogens except for F. Due to the very high electronegativity of F, HF is most polar, and also there is hydrogen bonding present in it. Hence, it has the second-highest melting point.

In Figure-8-5, it is observed that boiling points for group VI increases with increasing molar mass except for H_2O. Due to the presence of hydrogen bonding, H_2O has a higher boiling point than H_2S.

In Figure-8-6, it is observed that the boiling point increases with molar mass for hydrocarbons. All of these compounds are nonpolar and only have London dispersion forces, so the trend line fits closely to the curve of increasing boiling points with molar mass.

It was hypothesized at the beginning of this investigation that the melting or boiling points increase with increasing molar mass, which turns out to be partially true. There were some discrepant compounds like HF and H_2O due to the presence of hydrogen bonding. The results are observed in Figure-8-4. Figure-8-5, and Figure-8-6.

The real-world connection of intramolecular and intermolecular forces is the understanding of substances around us. The concept of intramolecular and intermolecular forces helps understand the states of matter, shapes of the molecule, the specific behavior of compounds, melting and boiling points, etc.

9 Lab 9: Reversible Reactions

Goal
To investigate interactions of an indicator with acids and bases to show reversible reactions.

Introduction

Reversible reactions are those chemical reactions, where the forward and backward reaction takes place. In the forward reaction, reactants react to form products, while in the backward reaction, the products of the reaction react together to produce the original reactants.

In the generalized form the reversible reaction is expressed as:

$$A + B \rightleftharpoons C + D$$

The symbol $\rightleftharpoons$ having two half arrowheads, one pointing in each direction, is used in equations that show reversible reactions. The forward reaction is the one that goes to the right, while the backward reaction is the one that goes to the left.

In a reversible reaction, the reaction mixture may contain reactants and products in certain proportions at any instant, which may be changed by altering the reaction conditions.

Example: Ammonium chloride breaks down when heated, forming ammonia and hydrogen chloride, while these two gases, when cold, react together to form ammonium chloride again.

This reversible reaction can be modeled as:

$$NH_4Cl(s) \rightleftharpoons NH_3(g) + HCl(g)$$

Energy change in a reversible reaction

In a reversible reaction, if a reaction is exothermic in one direction, it is endothermic in the other direction. At dynamic equilibrium, the same amount of energy is transferred in both the forward and reverse reactions.

Example:

Blue copper sulfate is generally hydrated. The copper ions in its crystal lattice structure are surrounded by water molecules. This water is driven off when blue hydrated copper sulfate is heated, leaving white anhydrous copper sulfate. The anhydrous copper sulfate, when cooled down, forms again hydrated copper sulfate. This reaction is expressed as:

$$CuSO_4 \cdot 5\,H_2O(s) \rightleftharpoons CuSO_4(s) + H_2O(l)$$

The forward reaction is endothermic and the reverse reaction is exothermic.

Dynamic Equilibrium

When a reversible reaction takes place in a closed container, it reaches a dynamic equilibrium. At equilibrium:

- the forward and backward reactions are continuously taking place

- the forward and backward reactions have the same rate of reaction
- the concentrations of all the reacting substances remain constant

Based on the concentrations of all the different reaction species at equilibrium, a quantity is defined called the equilibrium constant K_c, which is also sometimes written as K_{eq} or K. The equilibrium constant K_c describes the molar concentration in mole per liter at equilibrium for a specific temperature.

For a reversible reaction expressed as

$$aA + bB \rightleftharpoons cC + dD$$

and the value of equilibrium constant K_c is determined as

$$K_c = \frac{[C]^c [D]^d}{[A]^a [B]^b}$$

where *[C]* and *[D]* are equilibrium product concentrations; *[A]* and *[B]* are equilibrium reactant concentrations; and *a, b, c,* and *d* are the stoichiometric coefficients from the balanced reaction.

Le Chatelier's principle

Le Chatelier's principle states that changes in the temperature, pressure, volume, or concentration of a system at a particular dynamic equilibrium will result in predictable and opposing changes in the system in order to achieve a new equilibrium state. Hence, if a dynamic equilibrium is disturbed by changing the conditions, the position of equilibrium moves to counteract the change.

This principle can be used to predict the behavior of a system due to changes in pressure, temperature, or concentration.

Changes in Concentration

According to Le Chatelier's principle, adding additional reactant to a system will shift the equilibrium to the right, towards the side of the products. By the same logic, reducing the concentration of any product will also shift the equilibrium to the right. The position of equilibrium moves to the right.

$$A + 2B \rightleftharpoons C + D$$

If the reactant A is added to the system, the position of equilibrium moves so that the concentration of A decreases again - by reacting it with B and turning it into C + D. Hence, the concentration of A and B decrease, whereas C & D increase.

If the reactant A is removed from the system, the position of equilibrium moves so that the concentration of A increases again. That means that more C and D react to replace the A that has been removed. Hence, the position of equilibrium moves to the left.

Changes in Pressure

A change in pressure or volume will result in an attempt to restore equilibrium by creating more or fewer moles of gas. For example, if the pressure in a system increases or the volume decreases, the equilibrium will shift to favor the side of the reaction that involves fewer moles of gas. Similarly, if the volume of a system increases or the pressure decreases, the production of additional moles of gas will be favored.

Addition of an Inert Gas

If an inert gas, such as krypton or argon, is added to the reaction vessel, there will be no shift in the equilibrium position. The system always shifts so that the ratio of products and reactants remains equal to Kp or Kc. An inert gas does not react with either the reactants or the products, so it doesn't have any effect on the product/reactant ratio, and therefore, it has no effect on the equilibrium.

Changes in Temperature

For an endothermic reaction, heat is absorbed in the reaction, and the value of ΔH is positive. Raising the temperature on an endothermic reaction is essentially like adding more reactant to the system, and therefore, by Le Chatelier's principle, the equilibrium shifts the right. Conversely, lowering the temperature on an endothermic reaction shifts the equilibrium to the left, since lowering the temperature, in this case, is equivalent to removing a reactant.

For an exothermic reaction, heat is a product. Therefore, increasing the temperature will shift the equilibrium to the left, while decreasing the temperature will shift the equilibrium to the right.

Hypothesis and variables

Hypothesis: The bromophenol blue will act as a pH indicator turning the color of the solution on successive addition of acid and base.

Independent Variable: Successive addition of 0.1M HCl and 0.1M NaOH to the solution of 2 drops of bromophenol blue added with 10 drops of H_2O

Dependent Variable: Change in color of the solution

Controlled Variable: 2 drops of bromophenol blue added with 10 drops of H_2O in well-A2

Materials:

- 0.1 M Hydrochloric acid
- 0.1 M Sodium Hydroxide
- Bromophenol blue indicator
- Distilled water
- 24-well reaction plate
- Plastic toothpick

Procedure:

1- A 24-well reaction plate was taken, and 10 drops of distilled water were put into wells A1, A2, and A3. Thereafter, 2 drops of bromophenol blue were added to each well A1, A2, and A3.

2- One drop of 0.1M HCl was added to well A1 and mixed with a plastic toothpick.

3- The color of well A1 was recorded in Table-9-1.

4- Likewise, one drop of 0.1M NaOH was added to well A3 and mixed with an unused toothpick.

5- The color of well A3 was recorded in Table-9-1.

6- Two drops of 0.1M NaOH were added to well A1, mixed with the toothpick, and the color was recorded in Table-9-1.

7- Similarly, two drops of 0.1M HCl were added to well A3, mixed with the toothpick, and the color was recorded in Table-9-1.

8- Three drops of 0.1M HCl were added to well A1, mixed with the toothpick, and the color was recorded in Table-9-1.

9- Three drops of 0.1M NaOH were added to well A3, mixed with the toothpick, and the color was recorded in Table-9-1.

Results:

Table-9-1 records the amount of acid and base added successively to different wells and changes in the color of wells.

	Well-A1	Well-A2	Well-A3	Color-observed
1	10 drops of H_2O + 2drops of bromophenol blue	10 drops of H_2O + 2drops of bromophenol blue	10 drops of H_2O + 2drops of bromophenol blue	A1=Blue A2= Blue (No change) A3=Blue
2	1 drop of 0.1M HCl	No change	1 drop of 0.1M NaOH	A1=Yellow A2= Blue (No change) A3=Blue
3	2 drops of 0.1M NaOH	No change	2 drops of 0.1M HCl	A1=Blue A2= Blue (No change) A3=Yellow
4	3 drops of 0.1M HCl	No change	3 drops of 0.1M NaOH	A1=Yellow A2=Blue (No change) A3=Blue

Table-9-1: Observations of change in color for a reversible reaction

Discussion/Analysis

As observed in the experiment, bromophenol blue as a pH indicator has two colors, yellow and blue.

The color of bromophenol blue in acid is yellow. If the acid is diluted too much, it turns greenish color. Further dilution of acid turns the color of the solution blue. The color of bromophenol blue in a base is blue.

The color of bromophenol blue changes from blue to yellow by adding the acid HCl to the solution. Similarly, the color of bromophenol blue changes from yellow to blue by adding the base NaOH to the solution.

As shown in Image-9-1, the successive addition of acid and bases shifts the equilibrium of reversible reaction, which is indicated by a change in color. The bromophenol blue shows a yellow color when the solution is acidic, while it shows a blue color when the solution is alkaline. Bromophenol blue has a blue color when its pH is above 4.6, and has a yellow color when its pH is below 3.

For dilute acids having pH above 4.6, the color of bromophenol blue is blue. Bromophenol blue is not good for distinguishing weak acids and bases. Dilute acids, neutral, dilute base, and concentrated bases show blue color when mixed with bromophenol blue. Bromophenol blue shows greenish color for a transition pH range between 3.5 and 4.

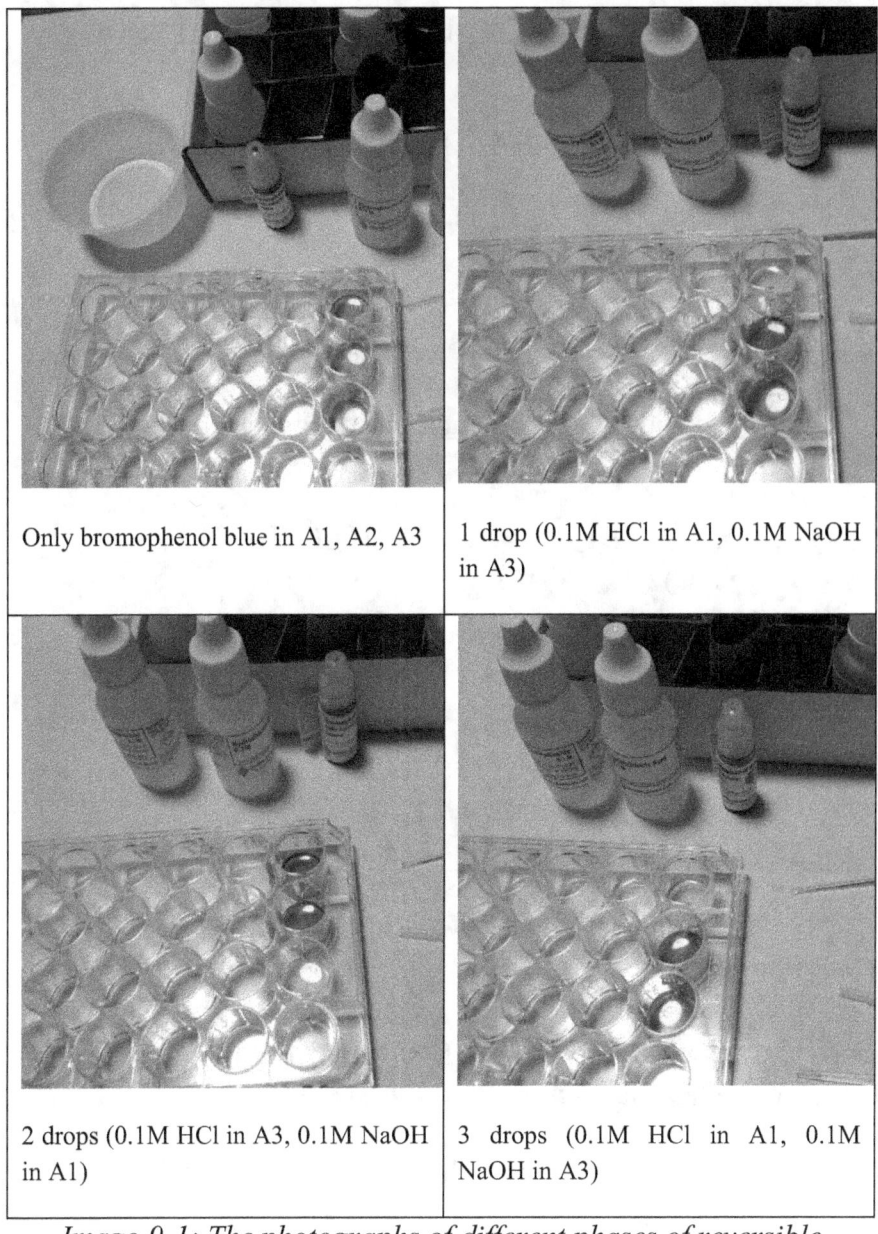

Image-9-1: The photographs of different phases of reversible reaction

Chemical equation for the reaction changing color from blue to yellow:

$$C_{19}H_{10}Br_4O_5S + H^+ \rightarrow C_{19}H_{11}Br_4O_5S^+$$

  (Blue)   (Yellow)

$$\text{-----------------------Eq(1)}$$

Chemical equation for the reaction changing color from yellow to blue:

$$C_{19}H_{11}Br_4O_5S^+ + OH^- \rightarrow C_{19}H_{10}Br_4O_5S + H_2O$$
  (yellow)   (Blue)

$$\text{-----------------------Eq(2)}$$

Application of Le Chatelier's principle in the experiment

Le Chatelier's principle applies to the system because the chemical reaction is reversible. If the chemical reaction in Eq(1) is considered as a forward reaction, then the chemical reaction in Eq(2) is a backward reaction. Combining Eq(1) and Eq(2), the complete reversible reaction is expressed as below:

$$C_{19}H_{10}Br_4O_5S + H^+ \rightleftharpoons C_{19}H_{11}Br_4O_5S^+ + OH^-$$

$$\text{-----------------------Eq(3)}$$

Equilibrium reactions are governed by Le Chatelier's Principle, which states that the system will shift in such a way as to reduce the stress placed on its position.

Adding hydrochloric acid, i.e. increasing the concentration of hydrogen ions present in solution, the system must shift in such a way as to decrease the concentration of hydrogen ions. Since the hydrogen ions are present on the reactants' side of the equation Eq(1), decreasing their concentration means shifting the equilibrium to the right.

This shows that the forward reaction, which is represented by Eq(1), is now being favored over the reverse reaction, which is represented by Eq(2).

Error mitigation

In this investigation, the equilibrium constants on the addition of acid and base were not calculated, which does not show the quantified shift in dynamic equilibrium. It would have been a better observation, had the equilibrium constants on every addition of acid and base been calculated.

Although observation of light and dark color might have given better estimates of the pH value of the solution, the color difference of light blue and dark blue were not distinguished in the experiment.

The transition of the pH range between 3.5 and 4 was not observed in the investigation. The successive amount of acid and bases added to the solution always increased by 1 drop. For better observation, the amount of addition should be less than 1 drop.

Bromophenol blue is not good for distinguishing weak acids and bases. Dilute acids, neutral, dilute base, and concentrated bases all show blue color when mixed with bromophenol blue.

Conclusion

Le Chatelier's principle states that when a system experiences a disturbance (such as concentration, temperature, or pressure changes), it will respond to restore a new equilibrium state. For example, if more reactants are added to a system, Le Chatelier's principle predicts that the reaction will generate more products to offset the change and restore equilibrium.

It has been observed that when a reaction at equilibrium is perturbed by applying a stress, the reaction responds by shifting its equilibrium position so as to counteract the effect of the perturbation/stress. When a reaction makes more products as a response to the perturbation, it is a right-shift expressed by Eq(1). When a reaction makes more reactants in response to the perturbation, it is a left-shift represented by Eq(2).

Adding hydrochloric acid, i.e. increasing the concentration of hydrogen ions present in solution, the system shifts in such a way as to decrease the concentration of hydrogen ions. Since the hydrogen ions are present on the reactants' side of the equation Eq(1), decreasing their concentration means shifting the equilibrium to the right. This shows that the forward reaction, which is represented by Eq(1), is now being favored over the reverse reaction, which is represented by Eq(2).

The reversible reaction in this investigation is modeled as

$$C_{19}H_{10}Br_4O_5S + H^+ \rightleftharpoons C_{19}H_{11}Br_4O_5S^+ + OH^-$$

It was observed that bromophenol blue acts as a pH indicator showing yellow color for the acidic solution and blue color for the alkaline solution. Bromophenol blue shows a yellow color for a pH less than 3. The transition pH range is between 3 and 4.6, for which bromophenol blue shows light greenish color. For high pH above 4.6, the bromophenol blue shows a blue color.

It was hypothesized at the beginning of this investigation that the bromophenol blue acts as a pH indicator turning the color of the solution on successive addition of acid and base, which turns out to be true, as observed in Table-9-1 and Image-9-1.

The real-world connection of reversible reactions is the understanding of a special type of reaction which dynamically proceeds in forward and reverse directions. The concepts of reversible reaction can be applied to address some sort of environmental problems we face today. In a naturally occurring reversible reaction, a harmful forward reaction can be slowed down or reversed by applying Le Chatelier's principle. There are several various real-world applications of Le Chatelier's principle that are used in daily life. Not only is Le Chatelier's principle mathematically related to the equilibrium constant, but it also helps in determining the equilibrium position of an object.

For instance, the dissolved CO_2 in blood is in equilibrium with protons and bicarbonate anion, where the amounts of each constituent at any given moment is influenced by Le Chatelier's principle and governed mainly by the inhaling and exhaling of carbon dioxide. If CO_2 levels are high, the equilibrium shifts toward the formation of more hydrogen and bicarbonate ions, producing a lower blood pH. If CO_2 levels are low, the shift is in the opposite direction, hydrogen ion concentration decreases and the pH goes up.

Lab 10: Acid-Base Titration, Buffer Solution

Goals
To study buffer solution
To study titration curves of acid and bases
To investigate the pH curve of a weak acid and strong base
To estimate ionization constant of a weak acid
To explore differences between a strong and weak acid

Introduction

Titration is a technique, wherein a solution of known concentration (titrant) is used to determine the concentration of an unknown solution (titrand or analyte).

Titration

In a titration process, titrant and analyte is a pair of acid and base, where the titrant is added slowly in successive steps to a known volume of the analyte until the reaction is complete. The volume of titrant added helps determine the concentration of the unknown analyte. An indicator is used to signal the end of the reaction, which is the endpoint. Acid-base titrations are monitored by the change of pH as titration progresses.

The equivalence point in a titration at which the amount of titrant added is just enough to completely neutralize the analyte solution. At the equivalence point in an acid-base titration, moles of base equal moles of acid, and the solution only contains salt and water.

A titration curve is a plot showing the pH values of the solution as the reagent is added into the analyte. The shape of the titration curve depends upon the strength of the analyte and titrant. There are four common cases:

Case-1) Analyte is a strong acid, and titrant is a strong base

Case-2) Analyte is a strong acid, and titrant is a weak base

Case-3) Analyte is a weak acid, and titrant is a weak base

Case-4) Analyte is a weak acid, and titrant is a strong base

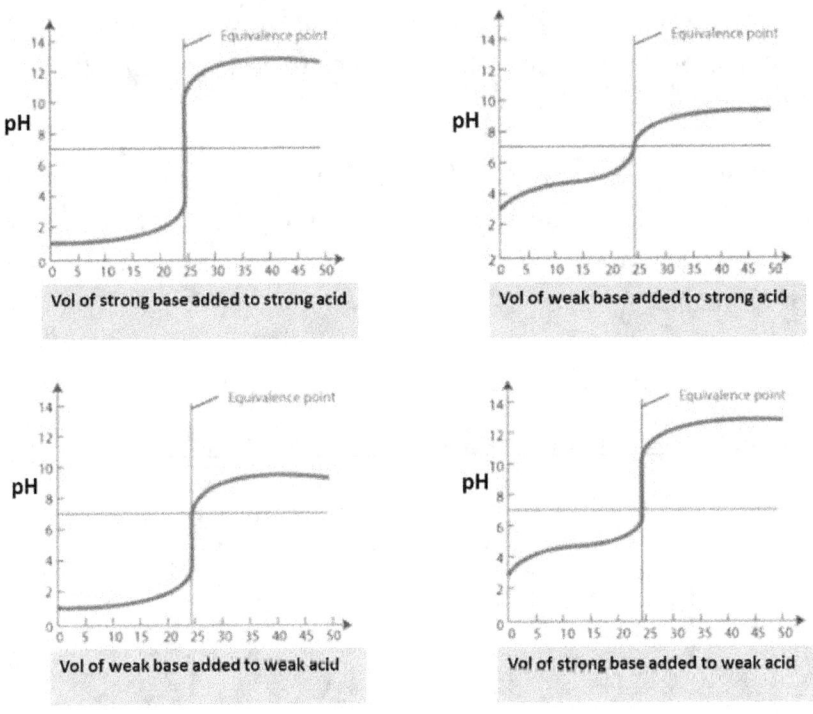

Figure-10-1: Titration curves for four important cases

Buffer solution

A buffer solution (also known as pH buffer or hydrogen ion buffer) is an aqueous solution consisting of a weak acid and its conjugate base, or vice versa. Its pH changes very little when a small amount of strong acid or base is added to it.

In an acid-base reaction, an acid and a base react to form a conjugate base plus a conjugate acid.

A conjugate acid is a chemical compound formed when an acid donates a proton (H^+) to a base. A conjugate acid contains one more hydrogen atom and one more positive charge than the base that formed it.

A conjugate base is what is left over after an acid has donated a proton during a chemical reaction. A conjugate base contains one less hydrogen atom and one more negative charge than the acid that formed it.

Example-1: Buffer solution from acetic acid and water

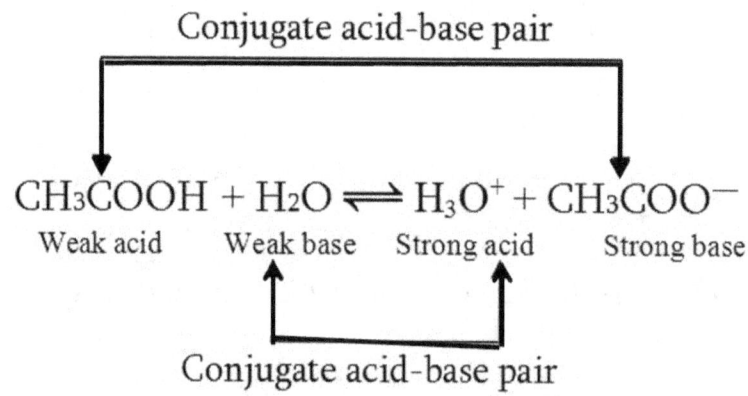

Figure-10-2: Conjugate acid-base pairs

In the figure above, the weak acetic acid (CH_3COOH) donates a proton to become a conjugate base, CH_3COO^-, of acetic acid.

$$CH_3COOH - H^+ \rightleftharpoons CH_3COO^-$$

In the reverse reaction, CH_3COO^- accepts a proton to form acetic acid, so CH_3COO^- is called a conjugate base.

Apart from that, H_2O behaves as a weak base, because it accepts a proton from acetic acid and becomes H_3O^+, which is the conjugate acid of H_2O.

$$H_2O + H^+ \rightleftharpoons H_3O^+$$

To sum up, the conjugate base is formed when an acid donates a proton, while conjugate acid is formed when a base accepts a proton.

Example-2: Buffer solution from hydroxide and ammonium ions

$$OH^- + NH_4^+ \rightleftharpoons H_2O + NH_3$$

In this example, the water molecule (H_2O) receives a proton donated by ammonium ion (NH_4^+), so water acts as the conjugate acid of the hydroxide ion. Similarly, ammonia is the conjugate base for the acid ammonium, as ammonium donates a hydrogen ion to produce the water molecule.

In the reverse reaction, the hydroxide ion acts as a conjugate base of water, since the water molecule donates a proton to produce an ammonium ion (NH_4^+). Similarly, ammonium ion acts as conjugate acid for the ammonia.

Buffer solutions are used as a means of keeping pH at a nearly constant value in a wide variety of chemical applications.

When a strong acid is added to an equilibrium mixture of the weak acid and its conjugate base, it implies that hydrogen ions (H+) are added, so the equilibrium is shifted to the left, in accordance with Le Chatelier's principle. Hence, the hydrogen ion concentration increases by less than the amount expected for the amount of strong acid added.

Similarly, if strong alkali is added to the mixture, the hydrogen ion concentration decreases by less than the amount expected for the amount of alkali added.

Ionization of strong/weak acids and bases

A strong acid dissociates (or ionizes) completely in an aqueous solution to form hydronium ions (H_3O^+).

$$HCl + H_2O \rightarrow H_3O^+ + Cl^-$$

A weak acid does not dissociate completely in an aqueous solution to form hydronium ions.

$$CH_3COOH + H_2O \rightleftharpoons CH_3COO^- + H_3O^+$$

A strong base dissociates completely in an aqueous solution to form hydroxide ions (OH^-).

$$NH_4OH \rightleftharpoons NH_4^+ + OH^-$$

Collection of strong acids/bases, and weak acids/bases:

Strong acids	**Weak acids**	**Strong bases**	**Weak bases**
Hydrochloric acid, Sulfuric acid, Nitric acid	Acetic acid, Hydrofluoric acid, Oxalic acid	Sodium hydroxide, Potassium hydroxide, Lithium hydroxide	Ammonium hydroxide, Ammonia

Table-10-1: Examples of strong acids, weak acids, strong bases, weak bases

Conjugate acid-base pairs

The conjugate base of acid is formed after an acid donates a proton. The conjugate acid of a base is formed after a base accepts a proton. A conjugate acid contains one more H atom and one more positive charge than the base that formed it. A conjugate base contains one less H atom and one more negative charge than the acid that formed it. The two species in a conjugate acid-base pair have the same molecular formula except the acid has an extra H^+ compared to the conjugate base.

Weak acids and weak bases always exist as conjugate acid-base pairs in an aqueous solution, expressed as:

$$\text{Acid} + \text{Base} \rightleftharpoons \text{Conjugate Base} + \text{Conjugate Acid}$$

The general chemical reaction showing acid and its conjugate base is:

$$HX + H_2O \rightleftharpoons X^- + H_3O^+ \quad \text{-----------Eq(1)}$$

Here, HX is the acid, and X^- is the conjugate base of HX. So, the anion (X^-) formed is a conjugate base of the acid.

Similarly, the general chemical reaction showing the base and its conjugate acid are:

$$X^- + H_2O \rightleftharpoons OH^- + HX \quad \text{-----------Eq(2)}$$

Here, X^- is a base and HX is the conjugate acid of X^-.

It is to be noted that weak acids have strong conjugate bases, while weak bases have strong conjugate acids. As shown in the above two reactions, if HX is a weak acid, then its conjugate base (X^-) will be a strong base. Similarly, if (X^-) is a weak base, then its conjugate acid HX will be a strong acid.

Changes in pH of buffer solutions

A buffer is an aqueous solution consisting of a mixture of a weak acid and its conjugate base, or a weak base and its conjugate acid. A buffer's pH changes very little when a small amount of strong acid or base is added to it. It is therefore used to prevent change in the pH of a solution upon the addition of another acid or base. A buffer's capacity is the pH range where it works as an effective buffer, preventing large changes in pH upon the addition of an acid or base.

- The pH of an effective buffer changes very little when a small amount of strong acid or base is added to it.

- The change in the pH of a buffer upon the addition of an acid or base can be calculated using the balanced equation and the formula for the equilibrium acid dissociation constant.
- Any buffer will lose its effectiveness if too much strong acid or base is added.

A buffer solution usually contains a weak acid and its conjugate base. When H^+ is added to a buffer, the weak acid's conjugate base accepts a proton (H^+), thereby "absorbing" the H^+ before the pH of the solution lowers significantly. Similarly, when OH^- is added, the weak acid donates a proton (H^+) to its conjugate base, thereby resisting any increase in pH before shifting to a new equilibrium point. In biological systems, buffers prevent the fluctuation of pH via processes that produce acid or base by-products to maintain an optimal pH.

pH, pKa, and Henderson-Hasselbalch Equation

The pKa is the pH value at which a chemical species accepts or donates a proton. The lower the pKa, the stronger the acid, and the greater the ability to donate a proton in an aqueous solution.

The Henderson-Hasselbalch equation relates pKa and pH as

$$pH = pKa + \log([\text{conjugate base}] / [\text{weak acid}])$$

For generalized example from Eq(1) and Eq(2), we have

$$pH = pKa + \log([X^-]/[HX])$$

pH is the sum of the pKa value and the log of the concentration of the conjugate base X⁻ divided by the concentration of the weak acid HX.

At half the equivalence point:

$$pH = pKa$$

Henderson-Hasselbalch equation is only an approximation, because it takes water chemistry out of the equation. This works when water is the solvent and is present in a very large proportion to the [H⁺] and acid/conjugate base. Hence, Henderson-Hasselbalch's equation should not be used for concentrated solutions, or for extremely low pH acids or high pH bases.

There are the following conditions required to apply Henderson-Hasselbalch's equation:

- $-1 < \log([X^-]/[HX]) < 1$
- Molarity of buffers should be 100x greater than that of the acid ionization constant Ka.
- Only use strong acids or strong bases if the pKa values fall between 5 and 9.

Hypothesis and variables

Hypothesis: At the equivalence point the solution will be basic i.e. pH greater than 7 for the titration of the weak acid (acetic acid) with the strong base (sodium hydroxide).

Independent Variable: Successive addition of 0.1M NaOH to the 50 drops (2.5 mL) of 0.1 M acetic acid.

Dependent Variable: pH value after each addition of 0.1M NaOH to 2.5 mL of 0.1M acetic acid

Controlled Variable: None

Materials

- 0.1M sodium hydroxide, NaOH
- 0.1M acetic acid, CH_3COOH
- 15 mL beaker
- All ranges of pH paper with at least 0.5 pH precision

Procedure:

1-The dropper bottle of 0.1M acetic acid was held vertically over a 15 mL beaker, and 50 drops were put in the beaker.

2-The pH paper strips were examined noting the range of each type.

3-The pH values were recorded in Table-10-2 each time for the 0.1 M solution of NaOH drops added to a beaker containing 0.1 M of acetic acid. The drops from the 0.1 M solution of NaOH were added as follows:

0, 10, 20, 30, 40, 45, 50, 55, 60, 70, 80, 90, 100.

4-At onset a piece of low range pH paper was taken. The end was dipped into the solution from step-1 for a second, the excess liquid was shaken off, and the color on the paper to the color chart was matched. The pH paper was kept in the solution for just a split second because the indicator might come off the paper otherwise.

5-It was continued to add 0.1M NaOH as listed in the Table-10-2. After each addition of NaOH, the pH values were recorded in Table-10-2.

6-There were three ranges of pH paper. As with the addition of NaOH, when the pH came close to the upper limit, the next range of pH paper was used.

7-The titration curve was plotted in Figure-10-3.

Results

Titration of 50 drops of 0.1M acetic acid with 0.1M NaOH is recorded inTable-10-2, where the corresponding pH value of the solution is recorded.

Sl	Drops of 0.1M NaOH added	pH
1	0	3.5
2	10	4.5
3	20	4.5
4	30	5
5	40	5.5
6	45	6
7	50	8.5
8	55	11
9	60	12
10	70	12.5
11	80	12.5
12	90	12.5
13	100	12.5

Table-10-2: Observation of pH values for successive addition of NaOH into acetic acid

Titration curve with equivalence point and half-equivalence point is shown in Figure-10-3

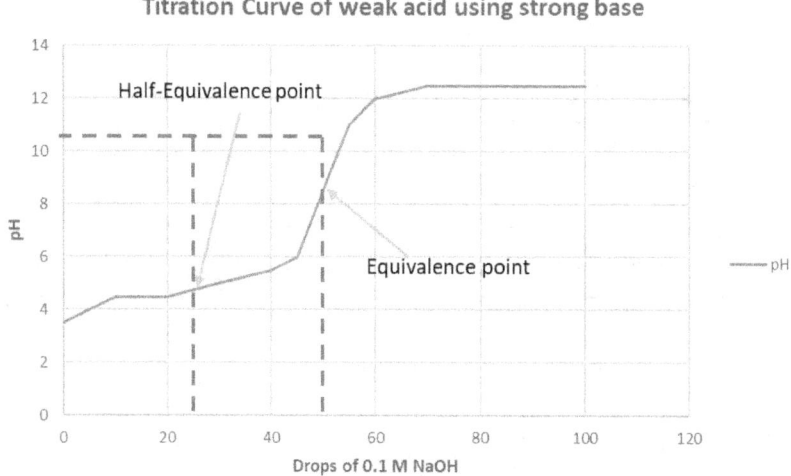

Figure-10-3: *Titration curve of a weak acid with a strong base*

The chemical reaction in the experiment taking place is expressed as:

$$CH_3COOH + OH^- \rightleftharpoons CH_3COO^- + H_2O$$

The titration curve in Figure-10-3 demonstrating the pH change during the titration of the strong base with a weak acid shows that at the beginning, the pH changes very slowly and gradually. This indicates the formation of a buffer system as the titration is around the half-equivalence point. At the equivalence point and beyond, the curve is very steeply increasing.

Discussion/Analysis:

Ka calculation using the Henderson-Hasselbalch equation

As observed in Figure-10-3, at equivalence point pH equals 8.5, and at half-equivalence point pH = 4.7.

The Henderson-Hasselbalch equation relates pKa and pH as

$$pH = pKa + \log([\text{conjugate base}] / [\text{weak acid}])$$

As per The Henderson-Hasselbalch equation, at half-equivalence point, pH = pKa

Using the pH of 4.7 at half-equivalence point, pKa = 4.7

$$pKa = -\log(K_a) = 4.7$$

$$K_a = 1.99 \times 10^{-5}$$

Observed Ka: 1.99×10^{-5}

Theoretical Ka = 1.8×10^{-5}

$$\% \text{ Error} = \left|\frac{Theoretical - Observed}{Theoretical}\right| \times 100\%$$

$$\% \text{ Error} = \left|\frac{1.8 \times 10^{-5} - 1.99 \times 10^{-5}}{1.8 \times 10^{-5}}\right| \times 100\% = 10.5\%$$

Comparison of titration-curve with different strength of analyte and titrant

A-Titration of a strong acid (HCl) with a strong base (NaOH):

1- At the start, no NaOH is added, so the pH of the analyte is low, as it predominantly contains H_3O^+ from the dissociation of HCl.

2-At equivalence point, moles of NaOH equals moles of HCl in the analyte. H_3O^+ ions are completely neutralized by OH^- ions. The solution is left only with salt and water. The pH is neutral i.e. 7.

3- Addition of NaOH continues, pH starts becoming basic, because HCl is completely neutralized and now excess of OH^- ions is available in solution.

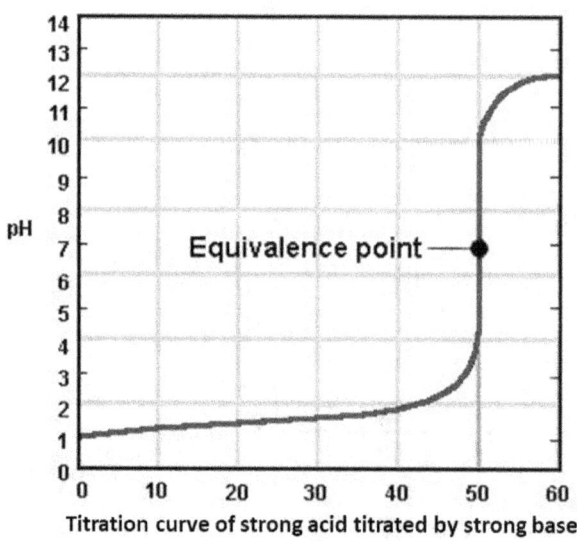

Titration curve of strong acid titrated by strong base

Figure-10-4: Titration curve of strong acid with a strong base

B-Titration of a weak acid (acetic acid CH_3COOH) with a strong base (NaOH):

1- At the start, there is no NaOH added, so the pH of the analyte is low, as it predominantly contains H_3O^+ from the dissociation of CH_3COOH.

2- As NaOH is added dropwise, H_3O^+ slowly starts getting consumed by OH^- ions. But analyte is still acidic due to the predominance of H_3O^+ ions.

3- At the equivalence point, moles of NaOH equals moles of CH_3COOH in the analyte. H_3O^+ is completely neutralized by OH^- ions. The solution contains only salt and water.

As weak acid has a strong conjugate base, so CH_3COO^- is a strong conjugate base of the weak acid CH_3COOH. CH_3COO^- reacts with water to produce hydroxide (OH^-) ions, thus increasing the pH above 7 at the equivalence point.

4- Beyond the equivalence point, when sodium hydroxide is in excess, the titration curve is identical to strong acid-strong base titration.

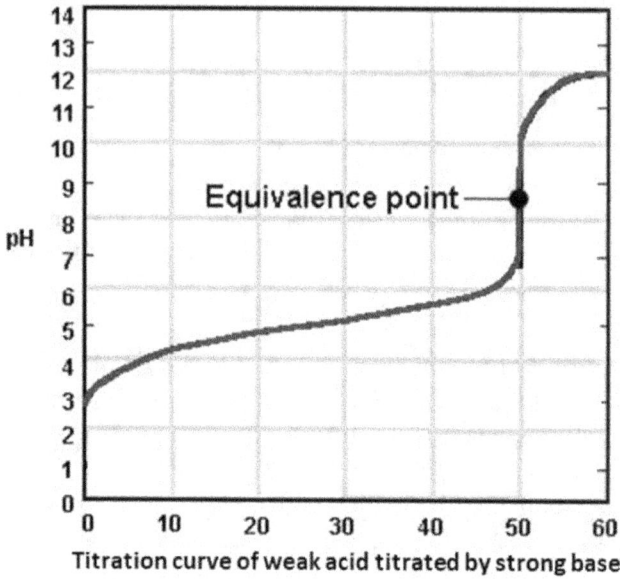

Figure-10-5: Titration curve of weak acid with a strong base

C-Titration of a strong acid (HCl) with a weak base (NH$_3$)

1- At the start, there is no NH$_3$ added, so the pH of the analyte is low, as it predominantly contains H$_3$O$^+$ from the dissociation of HCl.

2- As NH$_3$ is added dropwise, H$_3$O$^+$ slowly starts getting consumed by NH$_3$. The analyte is still acidic due to the predominance of H$_3$O$^+$ ions.

3- At the equivalence point, which is halfway up the steep curve, moles of NH$_3$ added equals moles of HCl in the analyte. The H$_3$O$^+$ are completely neutralized by NH$_3$.

4- In the case of a weak base versus a strong acid, the pH is not neutral at the equivalence point. The solution is acidic, pH < 7, at the equivalence point.

5- At the equivalence point, the solution only has ammonium ions NH_4^+ and chloride ions Cl^-.

6- The ammonium ion NH_4^+ is the conjugate acid of the weak base NH_3. So NH_4^+ is a relatively strong acid, and thus reacts with water to produce hydronium ions making the solution acidic.

7- After the equivalence point, NH_3 addition continues and is in excess, so the pH increases. NH_3 is a weak base so the pH is above 7, but is lower than compared with a strong base NaOH.

D- Titration of a week base (NH_3) with a week acid (CH_3COOH)

The analyte is NH_3 (a weak base) and the titrant is acetic acid CH_3COOH (a weak acid).

There isn't any steepness in the titration curve. There is just a point of inflection at the equivalence point. Lack of any steep change in pH throughout the titration renders titration of a weak base versus a weak acid ambiguous, and the titration curve does not convey much information.

Calculating theoretical and observed pH at different points of titration curve

A-Calculating starting pH value:

Acetic acid donates a proton to H₂O to form hydronium ions H_3O, and acetate ions CH_3COO^-.

$$CH_3COOH + H_2O \rightleftharpoons CH_3COO^- + H_3O^+$$

Our initial concentration of acetic acid is 0.1 molar. The initial concentration of acetate ions and hydronium ions is 0.

The change in concentration is inter-related. A loss of X concentration of acetic acid causes a gain of X concentration of acetate ions and X concentration of hydronium ions.

So, at equilibrium, a loss of X concentration of acetic acid implies

    (0.1 –X)   concentration of acetic acid

    X   concentration of acetate ions

    X   concentration of hydronium ions

Applying the principle of weak acid equilibrium, the equilibrium expression is

$$Ka = \frac{(X)(X)}{(0.1-X)}$$

Given $Ka = 1.8 \times 10^{-5}$

Solving for X, it is

    X = 0.00134

So, pH = $-\log[H_3O^+]$ = 2.87

Before any base is added to the acid, the pH of the titration curve should be 2.87, but in the observed titration curve the starting pH was 3.5.

B-Calculating pH at half-equivalence point:

Initially 50 drops (2.5 mL) of 0.1M acetic acid

At half equivalence point,

25 drops (1.25 mL or 0.00125 L) of 0.1 M NaOH implies (0.1 x 0.00125) i.e. 0.000125 moles of NaOH or OH^- ions

50 drops (2.5 mL or 0.0025 L) of 0.1M CH_3COOH implies (0.1 x 0.0025) i.e. 0.00025 moles of CH_3COOH.

The chemical reaction is expressed as:

$$CH_3COOH + OH^- \rightleftharpoons CH_3COO^- + H_2O$$

As seen, the OH^- ions act as a limiting reagent, and all of the moles of OH^- ions will be consumed in the reaction. So, 0.000125 moles of OH^- ions will be consumed and will neutralize 0.000125 moles of CH_3COOH.

Hence, 0.000125 moles of CH_3COOH will be left in the solution and 0.000125 moles of acetate ions CH_3COO^- will be gained.

The new volume of solution is 0.00375L from 25 drops (0.00125 L) of NaOH and 50 drops (0.0025 L) of acetic acid.

So, the concentration of acetic acid using moles/volume is

$$[CH_3COOH] = \frac{0.000125}{0.00375} = 0.0333 M$$

Similarly, the concentration of acetate ions is

$$[CH_3COO^-] = \frac{0.000125}{0.00375} = 0.0333 M$$

Hence, there is an equal concentration of a weak acid and its conjugate base, which shows that a buffer solution is formed.

The Henderson-Hasselbalch equation relates pKa and pH as

pH = pKa + log ([conjugate base]/[weak acid])

Since concentration of conjugate base i.e. acetate ions equals the concentration of weak acid i.e. acetic acid, so

pH = pKa

Given Ka = 1.8 x 10^{-5}, so pKa = 4.74

pH = 4.74

So, at the half-equivalence point the pH is 4.74, but in the observed titration curve the pH is 4.70.

C-Calculating pH at equivalence point:

At equivalence point,

50 drops (2.5 mL or 0.0025 L) of 0.1 M NaOH implies (0.1 x 0.0025) i.e. 0.00025 moles of NaOH or OH^- ions

50 drops (2.5 mL or 0.0025 L) of 0.1M CH_3COOH implies (0.1 x 0.0025) i.e. 0.00025 moles of CH_3COOH.

Moles is concentration times volume, so

moles of OH^- = (0.1 x 0.0025) = 0.00025 mol

Similarly, moles of CH_3COOH = (0.1 x 0.0025) = 0.00025 mol

The chemical reaction is expressed as:

$$CH_3COOH + OH^- \rightarrow CH_3COO^- + H_2O$$

There are the same amount of moles (0.00025 mol) of acetic acid and sodium hydroxide in the solution, so all of the acid is neutralized by the base. The solution is left without acetic acid and sodium hydroxide. After neutralization, 0.00025 mol of acetic acid is gained.

The total volume of solution is 5 mL or 0.005 L, where 2.5 mL comes from acetic acid and 2.5 mL from sodium hydroxide.

So, the molar concentration of acetate anions

$$[CH_3COO^-] = \frac{0.00025}{0.005} = 0.005M$$

The chemical reaction forming acetic acid and sodium hydroxide is expressed as

$$CH_3COO^- + H_2O \rightarrow CH_3COOH + OH^-$$

Whatever is lost from acetate is gained for acetic acid and sodium hydroxide. At equilibrium, the concentration for acetate is (0.005 - X), the concentration for acetic acid and sodium hydroxide is X.

So equilibrium expression for acetate acting as a base is

$$K_b = \frac{(X)(X)}{(0.005-X)}$$

Given, $K_a = 1.8 \times 10^{-5}$

$K_a \times K_b = 1.0 \times 10^{-14}$

So, $K_b = 5.6 \times 10^{-10}$

Solving for X, it is

$$X = \sqrt{(5.6 \times 10^{-10}) \times (0.005)} = 1.67 \times 10^{-6}$$

Hence, $[OH^-] = 1.67 \times 10^{-6}$

So, pOH = $-\log[OH^-]$ = 5.78

Since, pH + pOH = 14.00

Hence, pH = 8.22

As this is a titration of a weak acid with a strong base, so at the equivalence point, the pH is in the basic range, i.e. above 7. The theoretical pH at the equivalence point is 8.22, but in the observed titration curve the pH is 8.5.

Error mitigation

a-) In this investigation, the titration curve shown in Figure-10-3 is not smooth enough, because the volume of NaOH added in successive steps was in the bulk of 10 drops. For better observation of pH and thus better calculation of equilibrium expression Ka, the addition of NaOH should have been in the bulk of 5 drops or less than that.

b-) There could have been misjudging the color of the pH paper used as an indicator. The ranges of pH matching colors were very limited. For better observation of pH, more ranges of matching colors should have been provided.

c-) NaOH was added in drops for titration of acetic acid. Each drop is considered 0.05 mL. It could have been possible that each drop was not exactly the same as 0.05 mL, which might have caused some error in results. For better results, the volume should have been measured and added by a graduated cylinder.

Conclusion

In the experiment, the weak acid, acetic acid CH_3COOH, was neutralized adding small volumes in successive steps of the strong base, sodium hydroxide NaOH, to study the chemical phenomena of titration.

The titration of a weak acid with a strong base involves the direct transfer of protons from the weak acid to the hydroxide ion. The chemical reaction in the experiment taking place is expressed as:

$$CH_3COOH + OH^- \rightleftharpoons CH_3COO^- + H_2O$$

As shown in Figure-10-3, at equivalence point pH equals 8.5, and at half-equivalence point pH = 4.7.

A titration curve is a plot showing the change in pH of the solution in the conical flask as the reagent is added from the burette.

A titration curve can be used to determine:

1) The equivalence point of an acid-base reaction (the point at which the amounts of acid and base are just sufficient to cause complete neutralization).

2) The pH of the solution at the equivalence point is dependent on the strength of the acid and the strength of the base used in the titration.

-- For strong acid-strong base titration, pH = 7 at equivalence point

-- For weak acid-strong base titration, pH > 7 at equivalence point

-- For strong acid-weak base titration, pH < 7 at equivalence point

A titration curve in Figure-10-3 is the plot of the pH of the analyte solution versus the volume of the titrant added as the titration progresses. Figure-10-3 visually demonstrates the buffer capacity of the solution. The middle part of the curve near the half-equivalence point is flat because the addition of base or acid does not affect the pH of the solution drastically. This is the buffer zone, which contains a weak acid and its conjugate base. In this zone, the titration curve in the buffer zone is almost flat. When H^+ is added to a buffer, the weak acid's conjugate base accepts a proton (H^+), thereby absorbing the H^+ before the pH of the solution lowers significantly. Similarly, when OH^- is added, the weak acid donates a proton (H^+) to its conjugate base, thereby resisting any increase in pH before shifting to a new equilibrium point. However, once the curve extends out of the buffer region close to the equivalence point, it increases tremendously when a small amount of acid or base is added to the buffer system. If too much acid is added to the buffer, or if the concentration is too strong, extra protons remain free and the pH falls sharply.

In the experiment, the titration curve starts at pH 3.5, increases slowly and gradually with less steepness passing through the half-equivalence point (pH 4.7) till just before the equivalence point (pH 8.5), and then steeply increases after that.

In this investigation, there are some differences in actual and observed pH values. The titration curve shown in Figure-10-3 is not smooth enough, because the volume of NaOH added in successive steps was in the bulk of 10 drops. For better observation of pH, the addition of NaOH should have been in the bulk of 5 drops or less than that. Besides that, the ranges of pH matching colors were very limited. For better observation of pH, more ranges of matching colors should have been provided. Additionally, adding NaOH in drops, where each drop is considered as 0.05 mL in volume, might not have been correct, for which rather a graduated cylinder should have been used.

It was hypothesized at the beginning of this investigation that at the equivalence point the solution will be basic i.e. pH greater than 7 for the titration of the weak acid (acetic acid) with the strong base (sodium hydroxide), which turns out to be true, as determined in calculation besides observing in Table-10-2 and Figure-10-3.

The real-world connection of titration is in laboratory medicine to determine unknown concentrations of chemicals of interest in blood and urine. Pharmacists use titration in the development of new pharmaceuticals. The food industry uses titration to determine saturated fatty acids and unsaturated fatty acids. Often, titration is used to determine fat content, water content, and concentrations of vitamins. The production of biodiesel fuel is another area where titration is very commonly used. By measuring the pH of the biodiesel, one can easily determine how much base is needed to bring the solution to the correct pH, so that optimal fuel is created. Another common use of titration is testing water and soil. Plant growth depends upon the right condition of soil which is made optimal based on pH value determined by titration.

Lab 11: Redox Titration

Goals

To understand the concepts of redox titration
To determine the gain/loss of electrons by ions participating in redox titration

Introduction

A redox titration is a type of titration based on a redox reaction between the analyte and titrant. Redox Titration is commonly used as a laboratory method of determining the concentration of a given analyte. It is one of the most commonly used laboratory methods to identify the concentration of an unknown analyte. It may involve the use of a redox indicator and/or a potentiometer.

Redox titration is based on an oxidation-reduction reaction between the titrant and the analyte.

Reduction: A substance can undergo reduction via

- The addition of hydrogen.
- The removal of oxygen.
- The acceptance of electrons.
- A reduction in the overall oxidation state.

Oxidation: A substance can undergo oxidation via

- The addition of oxygen.
- Removal of hydrogen which was attached to the species.
- The donation/loss of electrons.
- An increase in the oxidation state exhibited by the substance.

Thus, it can be understood that redox titrations involve a transfer of electrons between the given analyte and the titrant. A common example of a redox titration is treating a solution of iodine with a reducing agent to produce iodide using a starch indicator to help detect the endpoint.

Key points:

- The titrant is the standardized solution; the analyte is the analyzed substance.
- Redox titration determines the concentration of an unknown solution (analyte) that contains an oxidizing or reducing agent.
- Not all titrations require an external indicator. Some titrants can serve as their own indicators, such as when potassium permanganate is titrated against a colorless analyte.

Lab-concept:

Potassium permanganate is used as an oxidizing agent to determine the hydrogen peroxide amount in a solution. Hydrogen peroxide reduces the permanganate to a colorless product. This redox reaction is performed in the lab.

Hypothesis and variables

Hypothesis: If the molarity of the potassium permanganate required for neutralization is observed in the experiment, then the oxidation state of the manganese product can be determined.

Independent Variable: Successive addition of 0.010 M $KMnO_4$ to the analyte.

Dependent Variable: Change in color as titrant is added to the analyte; Volume of 0.010 M $KMnO_4$ required for neutralization;

Controlled Variable: Room temperature.

Materials:

- 0.1 M HCl
- 0.010 M $KMnO_4$ solution
- 2 graduated pipettes
- 50 mL glass beaker
- Three beakers of size 15 mL, 30 mL, 150 mL
- Plastic funnel
- 2 Clear disposable cups
- Baking soda
- Antiseptic Hydrogen peroxide 3% (fresh bottle)
- Disposable plate
- Distilled water
- Digital balance
- Washing bottle

Procedure

Part1: Procedure - Calibration of the Pipets

1. Two graduated pipets were taken, where one was labeled with hydrogen peroxide and the other with potassium permanganate.

2. One cup was labeled with "distilled water" and filled with distilled water.

3. A clean, dry 15-mL plastic beaker was placed on the digital balance, and the balance was tared.

4. The densities of both potassium permanganate and hydrogen peroxide solutions were approximated as 1.00 g/mL, so that the volume (in mL) of each solution used in the titration could be readily determined.

5. One pipet was filled with distilled water and water was added dropwise to the 15-mL plastic beaker until 1.00 g of water had been added. The number of drops of water was recorded in Table-11-1.

Since the density of water and peroxide solution were considered 1.00 g/mL. So, drops/mL of water were the same as drops/mL of peroxide solution, thus recorded in Table-11-1. This pipet was reserved for use with the peroxide solution.

6. The cup was dried and repeated with the second graduated pipet, which was reserved for use with the potassium permanganate solution.

Part-2: Procedure - Preparation of the solutions on known concentration

1. The washing bottle is filled with distilled water. One of the disposable cups was labeled as "waste solution".

2. Antiseptic hydrogen peroxide used in this investigation was 3.0 % (w/w), which implied that 3.0 grams of hydrogen peroxide were available in 100 grams of solution. The antiseptic peroxide was approximately 80 times more concentrated than the 0.01 M permanganate solution. To simplify the observations of the titration, the concentration of hydrogen peroxide was diluted to make it nearly the same as the concentration of permanganate solution, so that the volumes of each solution used in the reaction was of the same magnitude.

A fresh bottle of hydrogen peroxide labeled 3% was used, because hydrogen peroxide slowly decomposes and older bottles generally have lower concentrations.

The clean, dried 15-mL plastic beaker was half-filled with hydrogen peroxide. To rinse and clean the graduated pipet, the pipet was half-filled with peroxide from the beaker, and then discarded into the waste solution cup.

3. A clean 150-mL plastic beaker was set on the balance tray and the balance was tared to zero. The peroxide graduated pipet was refilled with the peroxide solution from the 15-mL plastic beaker. The 3 % (w/w) peroxide solution was slowly added to the 150-mL plastic beaker on the balance until it was 2 g. The mass was recorded in Table-11-2.

4. The distilled water was added to make the total volume of 50 mL, for which the mass was also 50 grams. The washing bottle was used to add distilled water until the balance read 50.00 grams. The actual mass was recorded in Table-11-2. The beaker was

swirled to mix the solution.

5. The molar concentration of the mixture was determined for 2 grams of 3.0% hydrogen peroxide diluted to 50.00 mL.

$$\frac{3 \text{ g hygrogen peroxide}}{100 \text{ g Solution}} \times \frac{2 \text{ g solution}}{1} = 0.06 \text{ g } H_2O_2$$

$$\frac{0.06 \text{ g hygrogen peroxide}}{0.05 \text{ L Solution}} \times \frac{1 \text{ mol hydrogen peroxide}}{34.01 \text{ g hydrogen peroxide}} = 0.035 \text{ M } H_2O_2$$

The calculated concentration of the diluted peroxide solution is recorded in Table-11-2.

Part-3: Procedure - Redox Titration

1. The 30-mL plastic beaker was put on top of a white sheet of paper as a background to easily observe any color changes during the titration.

2. The peroxide graduated pipet was rinsed with the diluted peroxide solution from the 150-mL plastic beaker and then the rinsing solution was disposed of into the waste solution cup.
3. The graduated pipet was refilled with the diluted peroxide solution, and 30 drops of the solution were added to the 30-mL plastic beaker.

4. Using the 0.1 M HCl solution dropping bottle, 35 drops of hydrochloric acid were added to the glass beaker to acidify the peroxide solution.

5. The sealed potassium permanganate vial was placed on a disposable plate to protect the countertop from spills.

6. The sealed vial of 0.010 M potassium permanganate solution was opened.

7. The second pipet labeled for permanganate was filled with the permanganate solution from the vial.

8. Potassium permanganate was added dropwise to the acidified peroxide solution in the 30-mL plastic beaker on the white paper, counting the drops. The beaker was swirled after every couple of drops to mix the solution so that all the potassium permanganate could react.

9. As the endpoint was approached where all the peroxide had reacted, the solution stayed purple for a longer duration. As the solution approached the endpoint, only one drop at a time was added and swirled after each addition.

10. When the solution stayed a light-pink, the endpoint was reached. The number of drops of potassium permanganate solution needed to reach the endpoint was recorded in Table-11-3 in the unit of liter.

11. The titrated solution was disposed of in the waste solution cup.

12. The 30-mL plastic beaker was cleaned and steps 1 through 11 were repeated. The two trials were then averaged.

13. The remaining peroxide was poured from the beaker into the waste solution and peroxide was added until the waste solution is clear.

14. The clear waste solution was neutralized by adding baking soda until the solution stopped fizzing. Finally, the waste solution was disposed of.

Results

The data for the procedure of calibrating the pipets are recorded in Table-11-1.

	Number of drops	Mass (g)	Volume (mL)
Water	20	1	1
Hydrogen peroxide	20	1	1

Table-11-1: Data for number of drops, mass, and volume

The data for the procedure of making the solution on known concentration is recorded in Table-11-2.

Mass of 3% (w/w) Peroxide solution (g)	Total volume of diluted peroxide solution (L)	Concentration C of peroxide solution (M)	Moles of peroxide $N = C \times V$
2	0.05	0.035	0.00175

Table-11-2: Diluted peroxide solution

The data for the procedure of redox titration are recorded in the Table-11-3.

From the observed volume of the permanganate solution for two trials, the average volume is calculated. Additionally, the number of moles of permanganate is also calculated.

Number of moles (N) is related to concentration (C) and volume (V) as

$$N = C \times V$$

	Volume V of permanganate solution (L)	Concentration C of permanganate solution (M)	Number of moles N of permanganate $N = C \times V$
Trial 1	0.069	0.01	0.00069
Trial-2	0.070	0.01	0.00070
Average	0.0695	0.01	0.000695

Table-11-3: Amount of permanganate required for neutralization

Calculation/Analysis

Determining reduction of manganese

The possible oxidation states of manganese range from +7 through 0, where the most common states are +7, +6, and +2. Having recorded the observed data, the oxidation state of manganese is determined.

From the data of Table-11-2 and Table-11-3, it is observed that 0.000695 moles of permanganate (Table-11-3) react with 0.00175 moles of peroxide (Table-11-2).

When peroxide decomposes to oxygen, two moles of electrons are lost for one mol of peroxide.

$$O_2^{2-} \rightarrow O_2 + 2e^- \quad \text{-------------------Eq(1)}$$

Hence, for 0.00175 moles of peroxide oxidized in titration, there is a loss of 0.0035 moles of electrons.

In any redox reaction, moles of electrons lost equal moles of electrons gained. Therefore, potassium permanganate gains 0.0035 moles of electrons, while the moles of Mn^{7+} in the experiment is calculated as 0.000695.

The ratio of experimental moles of electron reduced and experimental moles of Mn^{7+} equals the ratio of moles of electron and moles of Mn^{7+}.

$$\frac{experimental\ mol\ of\ e^-\ reduced}{experimental\ mol\ of\ Mn^{7+}} = \frac{mol\ of\ e^-}{mol\ of\ Mn^{7+}}$$

Experimental mol of e^- reduced = 0.0035

Experimental mol of Mn^{7+} = 0.000695

Applying the formula the number of moles of electrons is determined for 1 mole of Mn^{7+}, so

$$moles\ of\ e^- = \frac{(0.0035)(1)}{(0.000695)} = 5.03$$

Hence, manganese ions (Mn^{7+}) gains 5 moles of electrons and gets reduced from +7 to +2, expressed as:

$$Mn^{7+} + 5e^- \rightarrow Mn^{2+} \quad \text{-----------------Eq(2)}$$

Determining balanced ionic equation

The unbalanced ionic reaction is

$$MnO_4^- + H_2O_2 + H^+ \rightarrow Mn^{2+} + 5O_2 \quad \text{---------------Eq(3)}$$

Simplifying with two half-reactions for redox reaction of Eq(3),

$$MnO_4^- + 8H^+ + 5e^- \rightarrow Mn^{2+} + 4H_2O \quad \text{-------------Eq(4)}$$

$$H_2O_2 \rightarrow O_2 + 2H^+ + 2e^- \quad \text{-------------Eq(5)}$$

Multiplying Eq(4) by 2, and Eq(5) by 5, and summing them up the balanced chemical reaction is

$$2MnO_4^- + 5H_2O_2 + 6H^+ \rightarrow 2Mn^{2+} + 5O_2 + 8H_2O \quad \text{------Eq(6)}$$

Validity check of experimental data

a- Ratio of moles of permanganate and peroxide

From the simplified balanced equation Eq(6), the ratio of moles of permanganate and peroxide is 0.4, which is used to validate the experimental data.

From Table-11-3, the number of moles of permanganate is 0.000695, and from Table-11-2, the number of moles of peroxide is 0.00175.

So, the experimental ratio of moles of permanganate and peroxide is $0.000695 / 0.00175 = 0.397$

The experimental ratio is very close to the theoretical ratio of moles of permanganate and peroxide, so it is verified that experimental data are correct.

b- Percentage error in volume of permanganate required for neutralization

From the balanced reaction of Eq(6), the number of moles of potassium permanganate ($N_{MnO_4^-}$) is related to the number of moles of hydrogen peroxide ($N_{H_2O_2}$) as

$$\frac{1}{2} N_{MnO_4^-} = \frac{1}{5} N_{H_2O_2}$$

Number of moles (N) is related with concentration (C) and volume (V) as

$$N = C \times V$$

Hence,

$$\frac{1}{2} (C_{MnO_4^-} \times V_{MnO_4^-}) = \frac{1}{5} (C_{H_2O_2} \times V_{H_2O_2})$$

Molarity of hydrogen peroxide $C_{H_2O_2}$ from Table-11-2 is 0.035M

Volume of hydrogen peroxide $V_{H_2O_2}$ from Table-11-2 is 50 mL

Molarity of potassium permanganate $C_{MnO_4^-}$ is 0.01 M

Volume of potassium permanganate $V_{MnO_4^-}$ is determined by titration. = 70 mL

From Table-11-2, the experimental volume of permanganate required for neutralization is 69.5 mL. The percentage error is calculated for observed and theoretical data.

$$\% \text{ Error} = \left| \frac{Theoretical - Observed}{Theoretical} \right| \times 100 \%$$

$$\% \text{ Error} = \left| \frac{70 - 69.5}{70} \right| \times 100 \% = 0.7\%$$

The % error is considerably small, so the observed data of the experiment is quite good.

Error mitigation

a-) The pipette should be rinsed out properly with the appropriate solution.

b-) As the solution approaches the endpoint, only one drop of permanganate at a time should be added and swirled after each addition.

c-) For better observation, either the volume and concentration of the analyte is increased, or the concentration of the titrant is decreased.

d-) Any trial which gives the deviation above 10% for the volume of permanganate from the previous trial should be ignored and a fresh trial should be performed.

Conclusion

Redox titration is based on an oxidation-reduction reaction, so it involves a transfer of electrons between the given analyte and the titrant.

In the experiment, a 0.010 M concentration of potassium permanganate is used to neutralize a 0.035 M concentration of hydrogen peroxide.

From the observed volume of the permanganate solution for neutralization, the number of moles of permanganate is calculated. From the data of Table-11-2 and Table-11-3, 0.000695 moles of permanganate (Table-11-3) reacts with 0.00175 moles of peroxide (Table-11-2).

When peroxide decomposes to oxygen, two moles of electrons are lost per mole of peroxide. Hence, for 0.00175 moles of peroxide oxidized in titration, there is a loss of 0.0035 moles of electrons.

In any redox reaction, moles of electrons lost equal moles of electrons gained. Therefore, potassium permanganate gains 0.0035 moles of electrons.

The ratio of experimental moles of electron reduced with experimental moles of Mn^{7+} equals 5. So, Manganese ion (Mn^{7+}) gains 5 moles of electrons and gets reduced from +7 to +2, expressed as:

$$Mn^{7+} + 5e^- \rightarrow Mn^{2+}$$

The balanced chemical reaction for redox titration is determined as

$$2MnO_4^- + 5H_2O_2 + 6H^+ \rightarrow 2Mn^{2+} + 5O_2 + 8H_2O$$

The experimental ratio is very close to the theoretical ratio of moles of permanganate and peroxide. The experimental volume of permanganate required for neutralization is 69.5 mL, while the theoretical value is 70 mL. The percentage error is calculated as 0.7%, which is considerably small, so the observed data of the experiment is quite good.

It was hypothesized at the beginning of this investigation that if the molarity of the potassium permanganate required for neutralization is observed in the experiment, then the oxidation state of manganese product can be determined., which turns out to be true, as observed in Eq(2).

The real-world connection of titration is to determine unknown concentrations of chemicals. Wines can be analyzed for sulfur dioxide using a standardized iodine solution as the titrant. In this case, starch is used as an indicator; a blue starch-iodine complex is formed in the presence of excess iodine, signaling the endpoint. Pharmacists use titration in the development of new pharmaceuticals. The food industry uses titration to determine saturated fatty acids and unsaturated fatty acids. Often, titration is used to determine fat content, water content, and concentrations of vitamins. Another example is the reduction of iodine to iodide by thiosulphate, again using starch as the indicator. This is essentially the reverse titration, called an iodometric titration.

Lab 12: Electrochemistry – Galvanic Cells

Goals
To understand the concept of galvanic cells
To determine the voltage generated by an electrochemical reaction

Introduction

A galvanic cell, also called a voltaic cell, converts chemical energy into electrical energy by an oxidation-reduction reaction.

Galvanic Cells provide the electron flow in a redox reaction to perform useful work. Such cells find common use as batteries, pH meters, and fuel cells. The setup of the cell requires that the oxidation and reduction half-reactions are connected by a wire and by a salt bridge or porous disk. Electrons flow through the wire creating an electrical current. The salt bridge or porous disk allows the passage of ions in the solution to maintain charge neutrality in each half-cell. Without the salt bridge, the solution in the anode compartment would become positively charged and the solution in the cathode compartment would become negatively charged, because of the charge imbalance, the electrode reaction would quickly come to a halt, therefore it helps to maintain the flow of electrons from the oxidation half-cell to a reduction half-cell.

The direction of the current in a cell is determined by the standard reduction potential of each half-cell. For a reaction to be spontaneous, the overall cell potential must be positive. Therefore, the half-reaction with the greater reduction potential will be a reduction and the other half-reaction will be an oxidation. The electrode in the oxidation half-reaction is called the anode. The electrode in the reduction half-reaction is called the cathode.

A galvanic cell consists of two reactions, one that is an oxidation reaction and one that is a reduction reaction. This is because an oxidation reaction produces free electrons, and a reduction reaction requires free electrons, so the electrons can transfer between the two reactions and do electrical work and produce electricity. Both oxidation and reduction reactions are called half-reactions because free electrons are present either before or after the reaction. There are certain locations in the galvanic cell designated for the half-reactions. The reduction reaction occurs in the cathode. The oxidation reaction occurs in the anode. For a pair of chemicals, one can be selected as the cathode and the other will be the anode. Both of the chemicals that are included in the chemical equations on the sides of those equations without free electrons are called electrodes and have a wire connecting them to transmit electrons and let them use their electricity to power some device.

The voltage in a circuit may be referred to as electromotive force, electrical potential, or simply as voltage. Volts equal joules per coulomb. The reduction potential (reduction voltage) of a particular electrode is usually assigned using a standard hydrogen electrode as the reference. But other substances, like zinc, can also be used as the standard reduction potential.

The cell potential is equal to the oxidation potential of the more reactive metal plus the reduction potential of the less reactive metal. Thus,

$$V_{oxidation} + V_{reduction} = V_{cell} \quad \text{-----------------Eq(1)}$$

In a cell, the electrode with the smallest reduction potential will be oxidized and the electrode with the largest reduction potential will be reduced.

Reduction potentials are often given in tables, from which the oxidation potential could be determined by reversing the equation and potential sign.

Hypothesis and variables

Hypothesis: An electrode having higher reduction potential acts as a cathode, while an electrode with a lower reduction potential acts as an anode. The cathode gets reduced, while the anode gets oxidized in an electrochemical reaction.

Independent Variable:
    Electrode copper with electrolytes $Cu(NO_3)_2$, electrode zinc with electrolyte $Zn(NO_3)_2$;
    Electrode nickel with electrolyte $Ni(NO_3)_2$, electrode zinc with electrolyte $Zn(NO_3)_2$;
    Electrolyte concentration; Composition of salt bridge;

Dependent Variable: voltage

Controlled Variable: Salt bridge parameters, temperature, resistivity of voltmeter, cable length.

Materials

- Reagents:
    - 0.1 M copper nitrate, $Cu(NO_3)_2$
    - 0.1 M nickel nitrate, $Ni(NO_3)_2$
    - 0.1 M zinc nitrate, $Zn(NO_3)_2$
    - 0.1 M sodium acetate, CH_3COONa

- Reagents:
    - Copper metal
    - Nickel metal

- Zinc metal

- 24-well reaction plate
- Chromatography paper
- Small beaker
- Fine sandpaper
- Plastic toothpicks
- Digital voltmeter with alligator clips

Procedure:

Procedure - Galvanic Cell-1:

1-A 24-well reaction plate was used to build a galvanic cell.

2-For the electrolytes, 15 drops of 0.1M $Cu(NO_3)_2$ was put in a well, and then 15 drops of 0.1M $Zn(NO_3)_2$ was put in an adjacent well.

3-Small strips of copper and zinc were cleaned with sandpaper.

4-Two strips about 3 mm wide by 40 mm long were cut from a small piece of chromatography paper.

5-About 15 drops of 0.1 M sodium acetate solution was put in an empty well away from the cells with electrolytes.

6-The chromatography strip was soaked in the CH_3COONa solution and pushed into the two wells of electrolytes using a plastic toothpick, forming a salt bridge between two half-cells.

7-The zinc strip was placed in the $Zn(NO_3)_2$ solution, and the copper strip in the $Cu(NO_3)_2$ solution.

8- With the metal strips affixed to the alligator clips, the voltmeter was set on.

9- The voltage was measured and recorded in Table-12-1.

Procedure - Galvanic Cell-2:

1-A 24-well reaction plate was used to build a galvanic cell.

2-For the electrolytes, 15 drops of 0.1M $Ni(NO_3)_2$ was put in a well, and then 15 drops of 0.1M $Zn(NO_3)_2$ was put in an adjacent well.

3-Small strips of nickel and zinc were cleaned with sandpaper.

4-Two strips about 3 mm wide by 40 mm long were cut from a small piece of chromatography paper.

5-About 15 drops of 0.1 M sodium acetate solution was put in an empty well away from the cells with electrolytes.

6-The chromatography strip was soaked in the CH_3COONa solution and pushed into the two wells of electrolytes using a plastic toothpick, forming a salt bridge between two half-cells.

7-The zinc strip was placed in the $Zn(NO_3)_2$ solution, and the nickel strip in the $Ni(NO_3)_2$ solution.

8- With the metal strips affixed to the alligator clips, the voltmeter was set on.

9- The voltage was measured and recorded in Table-12-1.

Procedure - Oxidation/reduction potential:

10-The oxidation or reduction potentials of Cu and Ni were determined based on the measured cell potentials and the known reduction potential for zinc as $-0.762V$.

11-Finally, the reduction potentials for each cell were determined.

Results:

The data of the experimental procedure are recorded in Table-12-1.

	Volt measured (V)
Galvanic-Cell-1(Cu, Zn)	1.09
Galvanic-Cell-2(Ni, Zn)	0.50

Table-12-1: Voltage measured for two galvanic cells

Discussion/Analysis:

For galvanic cell 1

$$Cu^{2+} (aq) + Zn(s) \leftrightarrow Cu(s) + Zn^{2+} (aq)$$

Copper is being reduced from a 2+ charge to a 0 charge thus it is gaining two electrons. Zinc is being oxidized from Zn charge 0 to a 2+ charge thus it is losing electrons. Oxidation happens at an anode, so the zinc electrode acts as the anode, while the copper electrode acts as the cathode.

From Eq(1)

$$V_{oxidation} + V_{reduction} = V_{cell}$$

The known reduction potential for zinc is –0.762 V. As the zinc electrode is the anode, so the oxidation potential of zinc equals 0.762 V. The V_{cell} from Table-12-1 is 1.09V.

Hence, the reduction potential of copper is

1.09 V – 0.762 V = 0.328 V

$$V_{cell} = 0.76 \text{ V} + 0.33 \text{ V} = 1.09 \text{ V}$$

For galvanic cell 2

$$Ni^{2+} \text{ (aq)} + Zn(s) \leftrightarrow Ni(s) + Zn^{2+} \text{ (aq)}$$

Nickel is being reduced from a 2+ charge to a 0 charge thus it is gaining two electrons. Zinc is being oxidized from Zn charge 0 to a 2+ charge thus it is losing electrons. Oxidation happens at an anode, so the zinc electrode is the anode, while the nickel electrode is the cathode.

The known reduction potential for zinc is –0.762V. As the zinc electrode is the anode, so the oxidation potential of zinc equals 0.762V. The V_{cell} from Table-12-1 is 0.5V.

Hence, the reduction potential of nickel is

0.5 V – 0.762 V = –0.262 V

	Anode (oxidation)	Cathode (reduction)
Galvanic-Cell-1(Cu, Zn)	Zn	Cu
Galvanic-Cell-2(Ni, Zn)	Zn	Ni

Table-12-2: Anode and cathode for two galvanic cells

The calculated reduction potentials are recorded in Table-12-3.

Reduction potential of zinc (V)	Reduction potential of copper (V)	Reduction potential of nickel (V)
−0.762	+0.328	−0.262

Table-12-3: Calculated reduction potential of copper, and nickel

The standard reduction potential of a cell can be determined by subtracting the standard reduction potential for the reaction occurring at the anode from the standard reduction potential for the reaction occurring at the cathode.

The standard reduction potential of cell-1 (SRP_{cell-1}) is calculated as

$$SRP_{cell-1} = SRP_{copper} - SRP_{zinc}$$

$$= (0.328V) - (-0.762V) = 1.09V$$

The standard reduction potential of cell-2 (SRP_{cell-2}) is calculated as

$$SRP_{cell-1} = SRP_{nickel} - SRP_{zinc}$$

$$= (-0.262V) - (-0.762V) = 0.500V$$

	Standard reduction potential of cell (V)
Galvanic-Cell-1(Cu, Zn)	1.09
Galvanic-Cell-2(Ni, Zn)	0.500

Table-12-4: Calculated reduction potential of two cells

Conceptual exercises

Conceptual exercise-1: A cell having potential of + 1.02 V has a magnesium anode and a zinc cathode. Calculate the reduction potential of magnesium.

As the zinc electrode acts as the cathode, reduction takes place at the zinc electrode, and oxidation takes place at the magnesium electrode.

From Eq(1), $\quad V_{oxidation} + V_{reduction} = V_{cell}$

Cell potential is 1.02 V

Reduction potential of zinc as –0.762 V

As oxidation takes place at the magnesium electrode, the oxidation potential of aluminum is calculated as

$$V_{oxidation} = 1.02V - (-0.762 \text{ V}) = 1.782 \text{ V}$$

Hence, the reduction potential of magnesium is –1.782 V

Conceptual exercise-2: A cell has one copper and one nickel electrode. Calculate the potential of the cell using experimental data.

Copper has a reduction potential of 0.328V, while nickel has a reduction potential of −0.262V. This implies that copper is more easily reduced than nickel. So, nickel is oxidized and copper is reduced. The oxidation potential of nickel is 0.262V.

From Eq(1), $\quad V_{oxidation} + V_{reduction} = V_{cell}$

Hence

$$V_{cell} = 0.262V + 0.328V = 0.59V$$

Calculating experimental error

The theoretical reduction potential of nickel is −0.25 V, and of copper is 0.34 V. The experimental value of reduction potential of nickel is −0.262 V, and of copper is 0.328 V.

The error percentage is calculated as

$$\% \text{ Error} = \left| \frac{Theoretical - Observed}{Theoretical} \right| \times 100\,\%$$

Calculating the percentage error for reduction potential of nickel:

$$\% \text{ Error} = \left| \frac{0.25 - 0.262}{0.25} \right| \times 100\,\% = 4.8\%$$

The % error is considerably small, so the observed and calculated data of the experiment are of high quality.

Calculating the percentage error for reduction potential of copper:

$$\% \text{ Error} = \left| \frac{0.340 - 0.328}{0.340} \right| \times 100\,\% = 3.5\%$$

Error mitigation

a-) The metal strip could be dull and might not have a pure metal surface. The metal surface should be cleaned properly with sand-paper.

b-) The alligator clips might not have been firmly attached to the metal strips so that there could be a bad electrical connection, leading to the fluctuation in voltage.

c-) If there is not enough solution in the well, the solution could not reach the salt bridge. So, the well should be around 70% full.

d-) Mixing of solutions in the reaction plate might change concentrations and cause inaccurate results. The electrolytes in the reaction plate should not overflow.

Conclusion:

A galvanic cell converts chemical energy into electrical energy by an oxidation-reduction reaction. The cell potential is equal to the oxidation potential of the more reactive metal plus the reduction potential of the less reactive metal. Thus,

$$V_{oxidation} + V_{reduction} = V_{cell}$$

It is observed that the electrode with the smallest reduction potential will be oxidized and the electrode with the largest reduction potential will be reduced. A metal having higher reduction potential becomes a cathode, where reduction takes place. Similarly, a metal having lower reduction potential becomes an anode, where oxidation takes place.

The reduction potentials of Cu and Ni are determined based on the measured cell potentials and the known reduction potential for zinc. Finally, the reduction potentials for each cell are determined.

Applying the Nernst equation to a simple electrochemical cell such as the Zn/Cu cell explains how the cell voltage varies as the reaction progresses and the concentrations of the dissolved ions change. As the same concentration of electrolytes were used in each of the half-cells, so concentration does not affect the cell potential in this investigation.

It was hypothesized at the beginning of this investigation that an electrode having higher reduction potential acts as cathode being reduced, while an electrode with lower reduction potential acts as anode being oxidized, which turns out to be true, as observed in Table-12-3 and Table-12-4.

The real-world connection of electrochemistry is its vast usage in industries. The principles of cells are used to make electrical batteries. A battery is the most common example of a voltaic cell in everyday life. Commercial batteries are galvanic cells that use solids or pastes as reactants to maximize the electrical output per unit mass. Most commonly, galvanic cells are used for:

- digital cameras (lithium batteries)
- digital watches (mercury/silver-oxide batteries)
- military applications (thermal batteries)
- hearing aids (silver-oxide batteries)
- cellphones (alkaline batteries)

13 Index

A
Acid-Base Titration   166-192

B
Bond Polarity   111-136
Buffer Solution   166-192

C
Chromatography   34-47
Colorimetry   70-92
Covalent Solids   9-19

E
Electrochemistry   211-223

G
Galvanic cell   211-223
Gravimetric Analysis   94-109

I
Intermolecular Forces   136-149
Ionic Solids   9-19

M
Metallic Solids   9-19
Mole Ratio   21-32
Molecular Polarity   111-136
Molecular Solids   9-19

P
Paper Chromatography   34-47

R
Reversible Reactions   151-164
Redox Titration   194-209

T
Titration Acid-Base   166-192
Titration Redox   194-209
Types of Chemical Reaction   49-68

V
Voltaic Cell   211-223

Kalisey Series

Delivering the highest quality academic materials

- **Saurya Singh**
Kalisey Academy Publication

Copyright © 2021, Author & Publisher. All rights reserved. No part of this publication may be reproduced in any form by print, photo print, microfilm or any other means without written permission.

For any further enquiries about the book and content, contact: **info@kalisey-softek.com**

www.ingramcontent.com/pod-product-compliance
Lightning Source LLC
Chambersburg PA
CBHW071355210526
45465CB00001B/109